MW00478354

Can't Chew the Leather Anymore

Musings on Wildlife Conservation in Yellowstone from a Broken-down Biologist

P. J. White

Edited by Robert A. Garrott

Can't Chew the Leather Anymore
Musings on Wildlife Conservation in Yellowstone from a Broken-down Biologist
All Rights Reserved.
Copyright © 2016 P.J. White
Edited by Robert A. Garrott
v2.0

P. J. White is the Branch Chief of Wildlife and Aquatic Resources at Yellowstone National Park.

The Yellowstone Association

ISBN: 978-0-934948-39-5

Library of Congress Control Number: 2016012362

> Names: White, P. J. (Patrick James) | Garrott, Robert A., editor.
> Title: Can't chew the leather anymore : musings on wildlife conservation in Yellowstone from a broken-down biologist / P.J. White ; edited by Robert A. Garrott ; graphic Design by Charissa Reid.
> Other titles: Wildlife conservation in Yellowstone from a broken-down biologist
> Description: Yellowstone National Park : The Yellowstone Association, 2016. | Includes bibliographical references. | Description based on print version record and CIP data provided by publisher; resource not viewed.
> Identifiers: LCCN 2016012362 (print) | LCCN 2016008492 (ebook) | ISBN 0934948429 () | ISBN 0934948399 (pbk. : alk. paper)
> Subjects: LCSH: Wildlife conservation--Yellowstone National Park. | Wildlife conservation--Social aspects--Yellowstone National Park | Wildlife conservation--Political aspects--Yellowstone National Park. | Wildlife management--Yellowstone National Park. | American bison--Conservation--Yellowstone National Park. | Grizzly bear--Conservation--Yellowstone National Park. | Yellowstone National Park.
> Classification: LCC QL84.22.Y4 (print) | LCC QL84.22.Y4 W45 2016 (ebook) | DDC 333.95/4160978752--dc23

PRINTED IN THE UNITED STATES OF AMERICA

For my dad

Contents

NPS Photo / Neal Hebert

MANAGEMENT

LEADERSHIP

Introduction

In 2015, I was challenged in a collegial way by Dr. Susan Clark from the Yale School of Forestry & Environmental Studies to define what exactly we were trying to preserve in Yellowstone National Park with regards to wildlife. She also remarked that it was important for biologists and managers working in the trenches, if you will, to share their deliberations, experiences, and struggles; lest history eventually reflect retrospective idealism rather than the more honest, imperfect realities. These seemingly simple promptings led to many hours of contemplation and, eventually, the writing of this book.

The title of the book refers to a saying I heard from time-to-time growing up in upstate New York, which was recently made famous by Al Pacino in the movie *Scent of a Woman*[1]. It aptly describes my current shabby state and, in combination with the subtitle, provides

---

[1]Scent of a Woman. 1992. Directed by Martin Brest. Distributed by Universal Pictures, Universal City, California.

an honest appraisal. I have spent more than 15 years working as a biologist in Yellowstone, currently as the leader of the Wildlife and Aquatic Resources Branch. The work is extremely rewarding and I'd like to believe that my coworkers and I improved the preservation and restoration of wildlife therein for the benefit and enjoyment of people.

As you might expect with almost 4 million visits per year, preserving wildlife in the park can be challenging at times. It can also be quite contentious because everyone has an opinion about Yellowstone. It truly is the people's park and folks are passionate about how it should be managed. Visitors and nearby residents alike have a wide variety of expectations and conflicting recommendations, and there is no shortage of political wrangling to complicate matters. Thus, at various times, I thought about calling this book My Life in Purgatory—Sometimes Hell Doesn't Seem So Bad. But these self-pitying moments of frustration are fleeting and, overall, I am having a great time. There are few, if any, better places to work as a wildlife biologist, and the passion with which my colleagues strive to preserve and restore wildlife is truly uplifting.

My objective with this book is to advance the conservation of wildlife in the Yellowstone area (Figure 1) by providing up-to-date information on the scientific, social, and political issues influencing their management, as well as some practical options for their effective conservation. My hope is the advice and lessons provided herein, which were attained through much effort and many tribulations, will help newer professionals and students of wildlife conservation avoid at least some of the pitfalls I encountered. Also, I hope

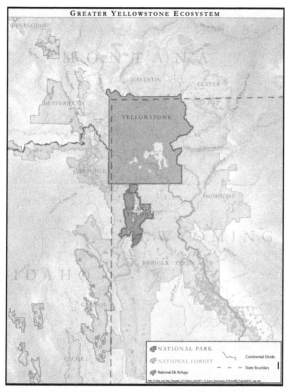

Figure 1. Map of the Greater Yellowstone Ecosystem.

this information will benefit the millions of people who visit the Yellowstone area each year or monitor the condition and management of the natural resources via the Internet or other outreach avenues.

The book is organized into three sections: philosophy, management, and leadership. Section I (philosophy) includes thoughts about ethics, expectations, goals, problem solving, and scientific knowledge. Essay 1 is about establishing achievable goals and solutions for the preservation of wildlife in and near Yellowstone. Essay 2 focuses on realistic expectations and the use of problem solving to restore native wildlife populations. Essay 3 provides cautions regarding underlying assumptions and the interpretation of acquired scientific knowledge. Essay 4 explores the ethics of killing wildlife that conflict with humans or native species restoration programs.

Section II (management) consists of case studies that illustrate the realities of wildlife conservation and management in and near Yellowstone. Essay 5 describes the challenges of managing migratory bison across jurisdictional boundaries, where different agencies have varying goals and philosophies. Essay 6 describes the challenges of preserving and restoring native fish given the park's history of promoting sport fishing and stocking nonnative fish. Essay 7 describes the recovery of grizzly bears and the challenges of managing them into an uncertain future given climate warming, conflicts with humans, and differing federal and state doctrines and expectations. Essay 8 explores the challenges of managing restored wolves given the divergence between reality and the symbolism of wolves for both advocates and opponents.

Section III (leadership) provides advice about knowledge, skills, and training related to interpersonal interactions that are necessary to be successful, but rarely discussed or taught in wildlife biology. Essay 9 provides tips on being an effective leader and promoting a culture of safety with your coworkers. Essay 10 discusses the importance of being an effective negotiator and striving for collaborative solutions rather than compromise. Enhancing wildlife conservation by building coalitions and developing feasible, practical solutions that lessen conflicts between humans and wildlife is a recurrent theme in the book.

NPS Photo / Neal Herbert

SECTION ONE:
Philosophy

Essay One:

Wildlife in Yellowstone: What Are We Trying to Preserve?

NPS Photo / Neal Herbert

Yellowstone National Park was established, in part, to "provide against the wanton destruction of the fish and game found within" for their "preservation in their natural condition ... for the benefit and enjoyment of people" (Park Protective Act of 1894). The park protects a diversity of aquatic, microbial, and terrestrial life in about 2.2 million acres of one of the largest temperate ecosystems in the world. Superintendents are tasked with conserving "the wild life therein ... by such means as will leave them unimpaired for the enjoyment of future generations" (Organic Act of 1916). As a result, about 90% of the park is currently managed as wilderness, where human disturbance is minimized compared to outlying areas and animals live in a relatively undisturbed setting. However, preserving natural conditions with minimal human impacts is challenging in modern society where rapid, widespread changes to habitats are occurring due to a warming climate, proliferating invasive species, and extensive development.[2] Likewise, preserving wild animals un-affected by human intrusion becomes more difficult as visitation to the park increases, and interactions and conflicts between people and wildlife increase.

Wildlife has been defined as animals living in a natural, undomesticated environment. The current management principles and wilderness ideals of the National Park Service strongly equate minimizing human intrusion with maintaining wildlife and wildness. However, it is well documented that native peoples significantly modified their environment for thousands of years through fires, harvests, and other activities. Also, the implementation of Yellowstone's mandate with regard to preserving wildlife has changed over time based on the prevailing attitudes, desires, and values of society at the time. Early in the park's history, many animals were harvested for food because there were few services for visitors. Hunting was prohibited in 1894, but angling continues to this day—though a catch-and-release philosophy now prevails. Until the 1960s, nature and wildlife were viewed as needing assistance or improvement from people. As a result, favorable nonnative fishes were introduced, predators and wild-fires were suppressed, ungulates were fed during some winters, hatchery-raised fish were stocked in lakes and rivers, and black and grizzly bears were allowed to beg along roads and feed at dumps.

This management paradigm was reversed during the 1970s and 1980s to reduce human intervention and allow natural ecological processes to prevail. This ap-

---

[2]Cole, D. N., and L. Yung, editors. 2010. Beyond naturalness: Rethinking park and wilderness stewardship in an era of rapid change. Island Press, Washington, D.C.

proach led to a wilderness ethic where humans were viewed as visitors rather than curators, and attempts were made to "re-wild" grizzly bears and other animals by removing human influences such as feeding. During subsequent decades, efforts were made to restore native species and the ecological processes that sustain them, including the recovery of bald eagles, gray wolves, grizzly bears, and peregrine falcons; the suppression of nonnative fish and the reintroduction of native cutthroat trout and Arctic grayling; and the recovery of a viable population of plains bison. To accomplish these activities, managers necessarily encroached on wildlife and wilderness principles in the short term to restore a more natural system in the long term.

Despite these efforts to promote naturalness and wildness, the intrusive management of wildlife still occurs in Yellowstone. Perhaps the best example is bison, which are intensely managed to limit their numbers and distribution when they migrate to the park boundary. This management occurs due to a 2000 agreement with the state of Montana designed to prevent transmission of the disease brucellosis from bison to cattle and limit the number of bison migrating outside the park. Since 1985, more than 6,600 bison have been culled from the population to protect the livestock industry, which has certainly hampered the recovery of bison in the ecosystem outside the park. In addition, some other animals (e.g., bears, coyotes, elk, wolves) are hazed or killed if they become dependent on human foods or attack or chase humans.

Today, visitors to Yellowstone have a wide variety of expectations regarding wildlife, with some wanting to see free-roaming, but approachable, animals that don't threaten their safety and others wanting unconstrained, uninhibited behavior by animals free of human influences. This contrast is reflected and reinforced by current management and messaging in the park. At entrance stations, visitors are warned wildlife are unpredictable and dangerous. Soon thereafter, however, they encounter large mammals near developed areas and along roadsides that are used to the daily presence of non-threatening people and, as a result, seem tame and safe to approach for viewing and photography. In these areas, rangers initially attempt to manage the behavior of people rather than wildlife to maintain separation and safety. However, animals are at times hazed from areas to alleviate traffic congestion and unsafe congregations of people venturing too close. For the relatively small portion of visitors who hike away from developed areas and roads, there are signs at trail heads to reinforce warnings about the unpredictable nature of wildlife and recommend safety measures. Also, there are required training videos in backcountry permit offices for visitors who venture overnight into wilderness areas. Wildlife are less accustomed to encountering people in these areas and, as a result, are generally less approachable and may react aggressively if surprised during unexpected close encounters (Table 1).

This wide range of historic and current perceptions and management approaches makes it difficult to discern precisely what managers in Yellowstone are trying to preserve with regards to wildlife. Clarity regarding what constitutes wildlife preserved unharmed in relatively natural conditions is important for evaluat-

Table 1. Management of the human-wildlife interface in Yellowstone National Park is divided into three broad zones, each with specific management strategies for visitor-wildlife interactions (Kerry Gunther, Bear Management Biologist, Yellowstone National Park).

Management Zones	Management Strategies
Developed Areas	• Managed for people to the exclusion of wildlife when conflicts occur. • Ungulates and small- to mid-sized mammals are tolerated. • Bears, cougars, and wolves are excluded. • Food-conditioned animals are removed. • Visitors are given priority when visitor and wildlife activities are not compatible.
Road Corridors	• Transition zone managed for both people and wildlife. • Managed for visitor transportation and wildlife viewing. • Animals are allowed to use roadside habitats for foraging and other natural behaviors. • Habituation of wildlife to people and people to wildlife is expected. • Food-conditioned animals are removed.
Wilderness/Backcountry	• Managed primarily for wildlife. • Overnight visitation is capped by a limited number of designated campsites. • Most human day-use is less than 3 miles from roads. • Implementation of seasonal recreational use closures for some bird nesting areas, wolf denning areas, and high use bear areas. • Animals are generally given priority in recreation management decisions where wildlife and human activities are not compatible. • Food-conditioned animals are removed.

ing whether the park's preservation mandate is being achieved, and also for deciding what intensity and types of management actions are appropriate in the park. Not too long ago, some colleagues and I proposed "a wild bison population can be defined as one that roams freely within a defined conservation area that is large and heterogeneous enough to sustain ecological processes such as migration and dispersal, has sufficient animals to mitigate the loss of existing genetic variation, and is subject to the forces of natural selection."[3] In hindsight, this technical description of an ecologically and genetically viable population did not

[3]White, P. J., R. L. Wallen, D. E. Hallac, and J. A. Jerrett, editors. 2015. Yellowstone bison—Conserving an American icon in modern society. The Yellowstone Association, Yellowstone National Park, Mammoth, Wyoming.

highlight a key component we are trying to maintain in animals—wildness, or untamed behavior.

While thinking about this oversight, I recalled some of the unrestrained behaviors and unspoiled conditions that evoked awe, fear, and wonder during my journeys in the park. I remember the warning "humph" of a grizzly bear after I had walked unaware to within 25 yards of him lying on a patch of snow in a narrow strip of trees on a high-elevation ridge—and breathlessly thanking him after he got up and walked away instead of attacking. I remember watching a very pregnant pronghorn doe pursue a fleeing coyote for more than one-half of a mile to protect a fawn or two that had yet to be born. I remember a peregrine falcon hurtling from the sky and chasing an aerobatic teal turn-for-turn until the duck eventually crashed into rocks along the river bank. I remember a grizzly bear almost effortlessly flipping a massive bison carcass over with one push, and a group of bison inexplicably banding together to defend a wounded elk from a pack of wolves for hours. And I remember finding an old female elk lying at the base of a tree during winter—having died in her sleep after living for about 20 years and likely producing many calves in an environment filled with predators and other dangers until her worn teeth failed her. These are a few of the memories that, to me, are the real expressions of wildness in Yellowstone.

Attempting to coalesce the essence of these recollections into a definition of "wildlife" is frustrating, however, and led me to recall conversations about Supreme Court Justice Potter Stewart who in 1964, when asked to define obscenity, indicated he couldn't define it, but he knew it when he saw it. So, how do we combine the technical and the intuitive to define a realistic bench-mark of what we are trying to preserve in Yellowstone with regards to wildlife? I propose the following: wildlife are untamed, free-roaming animals that live in an environment not dominated by humans and whose behaviors, movements, survival, and reproductive success are predominantly affected by their own daily decisions and natural selection. Mechanisms that contribute to natural selection include how wild animals compete with each other for food, mates, and space; how they protect themselves and respond to encounters with predators; who they choose to associate and mate with; and how they move about the landscape and endure variable environmental conditions. Animals with traits that make them better adapted to the environment will tend to survive, reproduce, and transmit their genetic characteristics to succeeding generations more than other animals. [4] Through this process, natural selection shapes wild behaviors and characteristics.

The word "untamed" in the definition incorporates the concept of wildness and uninhibited behavior, but is somewhat problematic because its meaning is subjective based on the perceptions of the observer and, as a result, could be difficult for managers to evaluate. Some people contend untamed animals should be wary of humans and/or avoid interactions with them, while others perceive untamed animals as free-roaming and unconfined, with their avoidance or indifference towards humans having little bearing on whether they are wild. I favor the latter classification because I don't equate a fear of humans with being wild. All animals

[4]Darwin, C. 1859. On the origin of species by means of natural selection. Down, Bromley, and Kent, London, United Kingdom.

develop their responses to humans based on experience. If these experiences are non-threatening, the animal learns there is no need to respond. Conversely, if the experiences are threatening or unpleasant in some way, the animals learn to flee and avoid areas frequented by people. As a colleague who conducts research in Antarctica points out, penguins on the continent aren't afraid of people and often approach with apparent curiosity, yet they are about as wild an animal as one could find.

Under my proposed definition for wildlife, actions such as feeding or husbandry that shield animals from forces of natural selection or attempt to mimic natural conditions in a zoo-like atmosphere should be avoided. Likewise, sustained or intensive human interventions such as repeated captures, handling, and culling should be avoided, if possible, because they could alter natural selection in unforeseen and unintended ways. While perhaps not optimal, animals habituated to human presence would still be considered wild if they maintained their independence from supplemental food and security provided by humans—despite close and frequent interactions.

At a time when wilderness appears to be less accessible to our increasingly urban society and, as a result, less important to most Americans, we should not underestimate the value of Yellowstone and the ease of access to its viewable geysers, scenery, and iconic wildlife for even the most urbanized citizens of the world. Many visitors (also known as voters) fondly remember their experiences at the park for the rest of their lives, even if many of the animals they observed were likely habituated to humans and the conditions that contributed to their inspiration were not pristine. It is unrealistic to expect that managers of Yellowstone National Park would, or even could, curtail visitation to the point where wildlife exist without any human influence. However, the majority of human intrusion through visitation can largely be confined to developments and road corridors where management is sometimes necessary to limit human-wildlife interactions and enhance human safety. Conversely, human intervention should continue to be limited in wilderness areas—though some intrusive actions will be necessary to restore native wildlife species and the processes that sustain them.

In all situations, wildlife management in the park should be based on reliable science (knowledge) and focus on maintaining viable populations of all native wildlife species, rather than protecting individual animals or merely iconic, large mammals. To the extent feasible, park rangers should continue to manage the behavior of people rather than wildlife to maintain separation and safety. Also, biologists and interpreters should discourage visitors from giving celebrity status to certain animals, which leads to their naming and anthropomorphism. Though some argue this practice helps connect people with nature, it also creates unrealistic expectations and issues for managers tasked with sustaining viable populations of wildlife rather than a zoo-like atmosphere where beloved individuals are guaranteed protection.

There are many key challenges facing wildlife and managers in the Yellowstone area during coming decades, and decisions need to be made regarding what refinements can and should be implemented to alleviate these stresses, if possible. As some of my friends and colleagues have witnessed, I can be quite opinionated and passionate about these topics. Therefore, I should probably start by admitting some of my major biases. I strongly believe one of our nation's finest ideals is that wildlife are considered a public resource and managed using reliable science for all Americans. I also maintain that some areas in this country, including Yellowstone, are so magnificent, unique, and important to our heritage that they should be preserved by the federal government for all citizens in perpetuity. Furthermore, I believe the long-term management of these iconic places should, to a large extent, reflect the will of the people across the country rather than being dictated solely by local interests, managers, or scientists. These entities should certainly be strongly represented, but they should not be allowed to compel decisions without adequate consideration of the science and national public input. This latter opinion is often difficult to apply because, at times, I find myself in the minority and society's attitudes and values change over time. However, I believe this principle is necessary to preserve the ideals for which these areas were created. As a result, preserve managers need to consider a wide variety of viewpoints and work diligently with adjacent communities, states, and the national public to reach reasonable solutions. This task is often extremely difficult in contemporary, contentious society, but it is essential because Yellowstone National Park is not large enough by itself to comprise a complete ecosystem for many wildlife species, which for their survival need to seasonally cross into other jurisdictions where they are subjected to different management approaches and philosophies.

The debate about how to conserve and manage wildlife in the Yellowstone area involves a variety of complex and controversial issues, including: How many are too many (or too few) animals for various species? Where and when will wide-ranging animals, particularly bison and large predators, be tolerated outside the park? How intensive should management be to minimize risks to property and human safety in local communities? What should be done and what can be done to suppress zoonotic diseases and/or lessen transmission risk to domestic animals and humans? What should be done to preserve existing genetic diversity and adaptive capabilities in threatened species? Should humans intervene to control ungulate numbers and grazing effects? And how, when, and where should hunting occur, while respecting tribal treaty rights and the concerns of other stakeholders?[5]

These questions can generally be categorized into the social sciences, ecosystem management, and pervasive human effects such as climate warming. Under the social sciences, it is unlikely visitation can be allowed to increase unfettered in forthcoming decades because we are already seeing adverse effects to wildlife, other park resources, and visitor experience. Also, the appropriate management of habituated wildlife (e.g., bears, elk) and visitors needs to be clarified in devel-

---

[5]White, P. J., R. L. Wallen, D. E. Hallac, and J. A. Jerrett, editors. 2015. Yellowstone bison—Conserving an American icon in modern society. The Yellowstone Association, Yellowstone National Park, Mammoth, Wyoming.

oped areas and along roadsides. Visitor education and safety messages need to be assessed and revamped, if necessary, because most visitors are relatively naïve when it comes to living with wildlife. Many people are not listening to warnings about the potential dangers of wildlife for reasons that include the cost of deterrents (e.g., bear spray), frequent turn-over of transient visitors, language barriers, and personal choice.

Furthermore, decisions need to be made regarding the current intolerance for certain species inside and outside park, with bats and bison being most indicative of this dilemma. Bats are decreasing at an alarming rate across much of North American due to white-nose syndrome and other factors, yet they are generally treated as pests rather than wildlife inside the park. Similarly, federal and state personnel prevent bison from accessing most public lands in the Greater Yellowstone Ecosystem due to persistent misperceptions about the risks of brucellosis transmission—especially when compared to other ungulates such as elk that are chronically exposed to brucellosis yet treated as wildlife and allowed to move freely through the system. The management of a viable population of 4,000 to 5,000 bison may not be sustainable in the long-term if they are prevented from migrating and required to forage year-round almost entirely in the park. Lastly, biologists need to do a better job of conveying why science matters, and that captures, experiments, flights, telemetry, and other intrusions are necessary to understand the system and answer uncertainties to help with preservation and restoration.

Under ecosystem management, there is a continuing need for collaborative and substantive dialogue between federal, state, and tribal representatives, which often have differing philosophies about how to manage wildlife, including the hunting of predators and ungulates adjacent to, and inside, the park. There is still a need for all of these parties to recognize and unite their somewhat different missions and management approaches to develop more effective, coordinated management of wildlife in the region. This is especially relevant given the recent recovery and expansion in distribution of large predators such as grizzly bears and wolves that are rejoiced by some people and despised by others. In addition, there needs to be further discussion about large-scale actions to restore native species and ecological processes, which involve relatively short-term human intrusions with the intent of long-term gain. However, many of these projects take decades to implement and have uncertain outcomes in terms of success and the eventual composition and condition of the restored system. As a result, such actions are expensive, time consuming, and controversial, even among park staff.

Lastly, we need to plan for the continual impacts of a warming climate, land use changes, and proliferating nonnative species and zoonotic diseases. Climate warming is already affecting precipitation and snowmelt patterns, river and stream flows, vegetation composition and phenology, and the frequencies of droughts and flooding, with observable effects on plant communities and water nesting birds such as trumpet-

er swans and white pelicans.[6]   It could also affect ungulate body condition and pregnancy, as well as grizzly
bear diets through decreased white-bark pine nuts and
cutthroat trout.   Continued habitat degradation and
fragmentation in the Greater Yellowstone Ecosystem
will have effects on wildlife migration and dispersal,
as well as foraging and other life history strategies. In
addition, increased invasions by nonnative species and
zoonotic diseases will threaten the adaptive capabilities and distributions of native species. These pervasive effects seem almost impossible to deal with effectively, but will likely be some of the primary influences
on wildlife and wilderness for the foreseeable future.

In the interest of self-preservation, I will not provide my ideas on how to address these challenges in
this essay. However, wildlife biologists, other scientists, and managers in Yellowstone have collectively
begun thinking about management refinements to
address them. In the interim, managers should continue to promote an environment where wildlife remain uncontrolled and visitors can be impressed and
inspired by their uninhibited behaviors. As historian
Paul Schullery emphasized, the greatest value of Yellowstone may be the "authenticity of its wildness—the
opportunity for us to be awed and learn from nature
making its own decisions."[7]

---

[6]Rodman, A., S. Haas, K. Bergum, C. Hendrix, J. Jerrett, N. Bow-
ersock, M. Gore, E. Oberg, L. Smith, and C. Reid. 2015. Ecological implications of climate change on the Greater Yellowstone
Ecosystem. Yellowstone Science 23:1-87.

[7]Schullery, P. 2010. Greater Yellowstone science: Past, present,
and future. Yellowstone Science 18:7-13.

Essay Two:

Wildlife Restoration: "You Can't Always Get What You Want"

It is quite possible Mick Jagger and Keith Richards were thinking about wild life when they penned these lyrics,[8] but I doubt they were of the Yellowstone variety. With apologies to the Rolling Stones, however, their famous refrain appropriately describes efforts to restore native wildlife species in the park. Prior to the 1970s, biologists attempted to improve nature and boost visitation by eliminating predators, stocking desirable fishes, allowing bears to feed at dumps and along roadsides, and culling ungulates due to perceived disease and overgrazing issues. Thereafter, biologists changed their approach and began allowing numbers of animals to fluctuate in response to competition, predation, resource availability, weather, and hunting when they migrated outside the park.[9] Biologists also initiated efforts to promote more natural conditions

---

[8] Jagger, M., and K. Richards. 1969. The Rolling Stones. Album "Let it Bleed." Decca, London, United Kingdom.

[9] National Research Council. 2002. Ecological dynamics on Yellowstone's northern range. National Academies Press, Washington, D.C.

by restoring viable populations of native species such as Arctic grayling, bald eagles, bighorn sheep, bison, black bears, common loons, cougars, elk, gray wolves, grizzly bears, peregrine falcons, pronghorn, trumpeter swans, and westslope and Yellowstone cutthroat trout. Managers faced a variety of obstacles in restoring these species, including biological uncertainties, logistical complexities, social intolerance, and persistent human conflicts. Some species such as bald eagles, cougars, elk, and pronghorn essentially restored themselves over time through increased survival and recruitment once overriding human stresses (DDT, persecution, culling, fencing) limiting their abundance and distribution were removed. Other species such as peregrine falcons initially required additional actions such as releases, nesting platforms, and less human disturbance to enhance recruitment. Extensive interdiction was necessary with black and grizzly bears to wean many of them from a dependence on human foods and reduce bear-human conflicts and human-caused deaths. Likewise, biologists had to remove nonnative fishes from several watersheds and install barriers that prevented reinvasion to enable the successful reintroduction of native Arctic grayling and cutthroat trout to some watersheds. Conversely, ongoing restoration efforts for bison and wolves largely involve dealing with social and economic issues, including conflicts with livestock, intolerance, and persistent misperceptions about disease transmission, property damage, and human safety. Other restoration actions remain incomplete despite intensive or prolonged attempts, including bighorn sheep hindered by recurring diseases originating from domestic sheep; an isolated, small population of common loons with poor dispersal and recruitment; the removal of millions of nonnative, predatory lake trout from Yellowstone Lake to restore cutthroat trout; and trumpeter swans with poor nesting success and recruitment due to climate warming (desiccated wetlands, flooding from rapid spring melt-off), human disturbance, and predation.

The outcomes of these bold, large-scale, restoration efforts will continue to unfold for many decades, but these actions have already provided several lessons to heed in the future. First, as Starker Leopold and colleagues lamented in 1963, "restoring the primitive scene [natural conditions] is not done easily nor can it be done completely ... exotic plants, animals, and diseases are here to stay."[10] This challenge has become even more difficult in recent decades with increasing impacts from development and visitation, invasive species, and warming temperatures.[11] Second, the maintenance or restoration of native species and ecological processes often requires that managers encroach on wildlife and wilderness principles in the short term to restore a more natural system in the long term. Monitoring and research are necessary for collecting information to effectively and efficiently accomplish restoration activities, as well as to determine the effectiveness of the actions. Also, complex projects such as the restoration of ecosystems and wa-

---

[10]Leopold, A. S., S. A. Cain, D. M. Cottam, I. N. Gabrielson, and T. L. Kimball. 1963. Wildlife management in the national parks. Transactions of the North American Wildlife and Natural Resources Conference 28:28-45.

[11]Cole, D. N., and L. Yung, editors. 2010. Beyond naturalness: Rethinking park and wilderness stewardship in an era of rapid change. Island Press, Washington, D.C.

tersheds may require intensive intervention and persistent maintenance actions over large areas for many decades. Moreover, restoration projects for different species may at times conflict, such as when loons are caught in nets set to remove nonnative lake trout or trumpeter swan cygnets are killed by restored bald eagles. Third, restoring predators or suppressing invasive nonnative species will not always undo the unwanted changes or issues caused by their prior elimination or introduction. In fact, restoring an altered system to its former condition is almost impossible because changes that occurred in the interim resist such a return.[12] Two examples that illustrate this latter issue include the restoration of large predators and attempts to restore cutthroat trout in Yellowstone Lake.

Cougars and wolves were eradicated from the park during the early 1900s, and numbers of bears were greatly reduced in the Yellowstone area. In turn, the abundance and distribution of elk increased in the absence of these top predators. Browsing by abundant elk reduced the establishment and survival of riparian shrubs and trees in northern Yellowstone, which along with widespread trapping by humans, contributed to a loss of beaver that require these trees for food and lodging and, in turn, a lowering of the water table as their dams disappeared.[13] As a result, the riverine sys-

---

[12]Hobbs, N. T., and D. J. Cooper. 2013. Have wolves restored riparian willows in northern Yellowstone? Pages 179-194 in P. J. White, R. A. Garrott, and G. E. Plumb, editors. Yellowstone's wildlife in transition, Harvard University Press, Cambridge, Massachusetts.

[13] Jonas, R. J. 1955. A population and ecological study of the beaver (*Castor canadensis*) of Yellowstone National Park. Thesis, University of Idaho, Moscow.

tem gradually transitioned over decades from riparian plant communities with beavers and tall willows to grasslands.[14] Wolves were released during 1995 to 1997 and rapidly increased in abundance and distribution, with elk their predominant prey. By 2003, there were about 100 wolves in northern Yellowstone, with concurrent increases in bear and cougar numbers, and continued liberal harvests of elk outside the park. As a result, elk numbers had decreased by more than 50% and the height and distribution of riparian species such as cottonwoods and willows were increasing in some areas.[15] Elk numbers continued to decrease to about 25% of their pre-wolf peak by 2012 due, in large part, to this diverse and abundant predator association.

There was certainly an expectation that the restoration of large predators to Yellowstone would eventually contribute to a substantial decrease in what many people perceived as overabundant elk in the northern portion of the park.[16] In turn, there was an expectation by many scientists that grasslands along rivers and streams would revert to the historic tall willow state as predators reduced elk browsing by substantially decreasing numbers and altering their foraging behavior. There were noticeable changes to the vegetation communities in some areas as elk numbers de-

[14] Meagher, M. M., and D. B. Houston. 1998. Yellowstone and the biology of time: Photographs across a century. University of Oklahoma Press, Norman, Oklahoma.

[15] Ripple, W. J., and R. L. Beschta. 2011. Trophic cascades in Yellowstone: The first 15 years after wolf reintroduction. Biological Conservation 145:205-213.

[16] National Research Council. 2002. Ecological dynamics on Yellowstone's northern range. National Academies Press, Washington, D.C.

creased by 75% with woody shrubs and trees growing taller and increasing, possibly due to less browsing by elk. However, the riverine system did not rapidly return to a widespread, tall willow state, and studies indicate other forces will be necessary to facilitate such a pervasive change, including the return of beavers and elevated water tables caused by their disturbances to water flowing through the system. These changes may take numerous decades to unfold because the abundant tall willows that beavers need for food and lodge building are no longer present and current water availability limits their establishment and survival.[17] It is encouraging that beavers have recently established colonies in some of the historic willow areas, but these changes are not yet occurring in other areas because of a recent quadrupling of bison numbers in northern Yellowstone due to high recruitment, survival, and dispersal from the central part of the park, possibly due in part to less competition from elk. This change has contributed to the continued intense browsing of riparian species in some areas.[18]

Another example of the difficulties of restoring an altered system to its original condition is the attempt to substantially suppress numbers of nonnative, predatory lake trout in Yellowstone Lake and restore cut-

[17]Hobbs, N. T., and D. J. Cooper. 2013. Have wolves restored riparian willows in northern Yellowstone? Pages 179-194 in P. J. White, R. A. Garrott, and G. E. Plumb, editors. Yellowstone's wildlife in transition, Harvard University Press, Cambridge, Massachusetts.

[18]Frank, D. A., R. L. Wallen, and P. J. White. 2013. Assessing the effects of climate change and wolf restoration on grassland processes. Pages 195-205 in P. J. White, R. A. Garrott, and G. E. Plumb. Yellowstone's wildlife in transition. Harvard University Press, Cambridge, Massachusetts.

throat trout. The Yellowstone Lake system developed over thousands of years without a top fish predator. Historically, millions of cutthroat trout primarily consumed zooplankton and amphipods, which in turn allowed more phytoplankton to grow due to less grazing. However, the system was transformed by the illegal introduction of predatory lake trout and parasitic whirling disease, which kills young fish and makes adults vulnerable to predation. Cutthroat trout were a major portion of lake trout diets following their introduction and, as a result, cutthroat numbers decreased from millions to hundreds of thousands.[19] In turn, the abundance of larger-bodied zooplankton increased and reduced the abundance of phytoplankton through grazing, thereby drastically altering conditions in the lake.[20] In addition, the collapse of the cutthroat trout population affected dozens of species that fed on them, including bald eagles, grizzly bears, osprey, otters, and

---

[19] Syslo, J. M. 2015. Dynamics of Yellowstone cutthroat trout and lake trout in the Yellowstone Lake ecosystem: A case study for the ecology and management of non-native fishes. Montana State University, Bozeman.

[20] Gresswell, R. E., and L. M. Tronstad. 2013. Altered processes and the demise of Yellowstone cutthroat trout in Yellowstone Lake. Pages 209-225 in P. J. White, R. A. Garrott, and G. E. Plumb. Yellowstone's wildlife in transition. Harvard University Press, Cambridge, Massachusetts.

pelicans, which could possibly have cascading effects through the ecosystem.[21, 22]

In an attempt to restore the system, the National Park Service has funded an intensive netting effort using park staff, commercial fishermen, and several large specialized netting boats. These efforts have killed about 1½ million lake trout since 2010 and predation on cutthroat trout is decreasing, with increased recruitment of juvenile cutthroat trout.[23] However, these costly, labor-intensive, and logistically complex efforts have not yet collapsed the lake trout population and likely cannot be sustained at current levels for many decades to come. Moreover, monitoring indicates the system may not be returning to historic conditions whereby cutthroat trout fed primarily on large-bodied zooplankton, but rather transitioning to a different state. After their release, lake trout fed heavily on

[21]Baril, L. M., D. W. Smith, T. Drummer, and T. M. Koel. Implications of cutthroat trout declines for breeding ospreys and bald eagles at Yellowstone Lake. 2013. Journal of Raptor Research 47:234-245.

[22]Middleton, A. D., T. A. Morrison, J. K. Fortin, M. J. Kauffman, C. T. Robbins, K. M. Proffitt, P. J. White, D. E. McWhirter, T. M. Koel, D. Brimeyer, and W. S. Fairbanks. 2013. Grizzly bears link non-native trout to migratory elk in Yellowstone. Proceedings of the Royal Society B 280:2013. 0870.

[23]Koel, T. M., J. L. Arnold, P. E. Bigelow, C. R. Detjens, P. D. Doepke, B. D. Ertel, and M. E. Ruhl. 2015. Native fish conservation program, Yellowstone fisheries & aquatic sciences 2012-2014, Yellowstone National Park. Report YCR-2015-01, National Park Service, Yellowstone Center for Resources, Yellowstone National Park, Wyoming.

cutthroat trout and their numbers and distribution proliferated. Soon thereafter, the cutthroat population collapsed and, with less cutthroat feeding on them, numbers of amphipods increased.[24] In response, both lake trout and cutthroat trout have shifted to primarily feeding on grazing amphipods. Thus, in addition to predation, cutthroat trout may eventually face competition for food from lake trout if amphipods become limiting.[24]

These two ecosystem-scale projects highlight that restoration takes a lot of time and, since we are still in the early stages of recovery, the eventual conditions of the restored systems remain uncertain. Therefore, the most interesting question moving forward is how conditions in these systems will continue to change. These examples also reiterate that it is unrealistic to expect we can return systems entirely to their historic, pristine conditions. In many cases, however, we can probably alter conditions to a more natural state that is much closer to desired conditions. As Leopold and colleagues (1963) indicated, "if the goal cannot be fully achieved it can be approached ... using the utmost in skill, judgment, and ecologic sensitivity."[25] Therefore, managers should specify feasible, realistic objectives for restoration actions, including the species and processes to be restored and the acceptable states or sets of

[24]Syslo, J. M. 2015. Dynamics of Yellowstone cutthroat trout and lake trout in the Yellowstone Lake ecosystem: A case study for the ecology and management of non-native fishes. Montana State University, Bozeman.

[25]Leopold, A. S., S. A. Cain, D. M. Cottam, I. N. Gabrielson, and T. L. Kimball. 1963. Wildlife management in the national parks. Transactions of the North American Wildlife and Natural Resources Conference 28:28-45.

conditions that could sustain them.[26] There will always be surprises, but by restoring native components and ecosystem processes scientists can improve the integrity and resilience of these systems. In the words of the Rolling Stones, "if you try sometimes, well you just might find, you get what you need."[27]

---

[26] Hobbs, R. J., D. N. Cole, L. Yung, E. S. Zavaleta, G. H. Aplet, F. S. Chapin III, P. B. Landres, D. J. Parsons, N. L. Stephenson, P. S. White, D. M. Graber, E. S. Higgs, C. I. Millar, J. M. Randall, K. A. Tonnessen, and S. Woodley. 2010. Guiding concepts for park and wilderness stewardship in an era of global environmental change. Frontiers in Ecology and Environment 8:483-490.

[27] Jagger, M., and K. Richards. 1969. The Rolling Stones. Album "Let it Bleed." Decca, London, United Kingdom.

NPS Photo/Daniel Stahler

Essay Three:

Best Available Science: Truth is in the Eye of the Beholder

NPS Photo/Neal Herbert

I often hear advocates, managers, and scientists refer to the "best available scientific information" when discussing the conservation and management of wildlife. I've done it many times myself. The phrase is generally used to refer to the most accurate, reliable, and pertinent information available for consideration in making a decision.[28] The general idea is that biologists implement rigorous studies to evaluate alternate explanations regarding the causes of observed phenomena or the effects of management actions. The quality of this work and any deductions are then reviewed anonymously and critically by several independent scientists (i.e., the peer-review process) to provide the best available science. The more studies that reach similar conclusions, the more reliable and verifiable the inferences are deemed to be.

---

[28]Ferrell, J. 2005. "Best available science" defined in proposed Endangered Species Act legislation. Marten Law, Seattle, Washington. http://www.martenlaw.com/newsletter/20051221-best-avail-science.

This paradigm has been extremely successful and, over time, has led to impressive advancements in knowledge about wildlife. As a result, attaining the best available science is generally embraced by scientists and the public alike. However, this lofty goal can also be problematic because science seldom produces the certainty in knowledge we seek for management or policy decisions in the short term. Rather, it can take a long time from the identification of an issue or uncertainty that needs resolution until the scientific process leads to a conclusive determination of cause and effect that is repeatedly verified and generally accepted. Furthermore, scientific investigation is often messy, with conflicting or disparate findings from different studies of the same issue. Science rarely results in clear, indisputable evidence that creates immediate, widespread consensus on the issue or process under investigation. It is hard to understand complex ecosystem processes and unequivocally identify the precise causes of past changes or forecast the consequences of proposed actions with a reasonable amount of certainty.

Because scientific investigation is often messy and slow to reveal the truth, there is an opportunity for people to knowingly or unwittingly support the interim or preliminary findings they like. We are all biased by our attitudes, experiences, and alliances. As a result, what is considered the best available science often depends on the interpretation or "professional judgement" of the beholder, which derives from each of our beliefs and values. For example, I recently reviewed comments on the proposed alternatives for a new bison management plan. Dozens of commenters alluded to the best available science in recommending a particular management approach. However, what each person considered the best available science generally varied depending on the stakeholder group with whom they were allied (e.g., animal welfare, livestock production, sporting, wildlife conservation).

Some advocates and other stakeholders selectively pick and choose scientific findings that support their beliefs, ethics, ideals, opinions, philosophy, and/or political agenda. They review the contrasting conclusions from various studies and selectively pick those most closely aligned with their own views. They then market or promote these findings as the best available science in an attempt to change or influence management and policy decisions. All debates about natural resource issues and policies have a strong component of human values underlying the science. Thus, even the most accurate and precise scientific information will rarely change the underlying values of advocates and segments of society with a predisposed stake in decisions regarding how our resources should be managed. In addition, the politicization of science is quite common these days, with more than a few people and groups having agendas and selectively using findings from scientific studies to advocate for their particular viewpoints and misrepresent the oppositions'. If the eventual decision goes against their beliefs, these advocates often claim the decision was based on poor science and demand additional scientific review and reconsideration.[29] Under these circumstances, the phrase "best available science" has little to no meaning.

Each scientist brings their own background, perspectives, and training to an investigation. These experiences and viewpoints influence how a study is designed and how the data are interpreted. As a result, equally well-trained scientists can evaluate the same data and arrive at different conclusions. In turn, conflicting interpretations from different scientific studies, all rigorously conducted and peer-reviewed, are often published in well-respected scientific journals. It is not surprising that various researchers make different interpretations when analyzing similar sets of complicated data. However, confusing and disparate findings, along with incomplete information, can frustrate managers and pose a dilemma for them when making decisions. If biologists don't agree on the findings or interpretation of existing research on the same topic, managers are more likely to base their decisions on values, politics, and other pressures. Inadvertently, they may choose the available science that supports their preconceived notions since it is human nature to prioritize and use what we inherently believe over other information. Also, a lack of consensus among scientists can be exploited by competing advocacy or political groups, with everyone claiming their position on some resource policy is most supported by the best available science because it is often possible to

[29]Sullivan, P. J., J. M. Acheson, P. L. Angermeier, T. Faast, J. Flemma, C. M. Jones, E. E. Knudsen, T. J. Minello, D. H. Secor, R. Wunderlich, and B. A. Zanetell. 2006. Defining and implementing best available science for fisheries and environmental science, policy, and management. Fisheries 31:460-465.

craft equally compelling arguments for opposing viewpoints based on the selective use of available science.

In addition, the complexities of ecological systems and scientific investigation can lead to contentious and longstanding debates among scientists about what or whose science is best. Like most people, biologists tend to gravitate towards others with similar beliefs and values. These associations of like-minded people are great for synergizing ideas and efforts to maximize productivity, and variety can generate constructive, critical debates about study design and analyses and interpretations of data. However, cliques or coteries can also unconsciously lead to group think and intolerance for, or at least less consideration of, alternate viewpoints. In addition, there is quite a bit of competition among these coalitions, which can exert itself during the grant request and peer-review processes. At its worst, this competition can contribute to a race to be first, with preliminary findings with minimal or tenuous support being reported as pervasive and irrefutable effects. Some of these claims become dogma in the minds of the public well before robust, reinforcing evidence is obtained to support them, and without regard for evidence countering their validity.

So what is the best available science? Well, it depends on your perspective and desires for the natural resource issue under consideration. The collection, evaluation, and dissemination of scientific information can be biased by our human shortcomings and tendencies more than we sometimes care to admit. One could argue this makes the phrase "best available science" shallow and useless. Rigorous scientific information is certainly essential for effective wildlife conservation. However, the way it is selectively marketed and promoted in contentious policy debates by various advocates, stakeholders, and some scientists erodes society's confidence in the credibility and utility of science. People become skeptical when they see and hear that science is not providing the answers needed to direct management or set policies because the participants in the debate often disagree on what constitutes the best available science. Therefore, we need to be mindful of these pitfalls and shortcomings as we collect and evaluate rigorous scientific information to make wise decisions. Also, we need to be transparent about our underlying values and the uncertainties in available information. Furthermore, we need to provide the rationale for decisions based on this information.[30] Debate is essential in the scientific process, but as my friends used to quip growing up, "consider the source."

---

[30]Sullivan, P. J., J. M. Acheson, P. L. Angermeier, T. Faast, J. Flemma, C. M. Jones, E. E. Knudsen, T. J. Minello, D. H. Secor, R. Wunderlich, and B. A. Zanetell. 2006. Defining and implementing best available science for fisheries and environmental science, policy, and management. Fisheries 31:460-465.

NPS Photo/Diane Renkin

Essay Four:

Killing in the Name of Conservation: The Death of Blaze

"You're not bear management, you're bear murderers. You fucking pieces of shit. I hope you all get eaten by bears. You miserable, rotten, horrible, shitty, fucking people. Why didn't you just at least send the bear to a sanctuary? Why did you have to kill her. YOU SUCK! You are horrible fucking people. How do you sleep at night? You disgusting pieces of shit. Die! I hate you! My tax money killed that bear. What is wrong with you? What is wrong with you?"

This anonymous voice message left for the manager of Yellowstone's bear program leaves you with a warm and fuzzy feeling, doesn't it?[31] Dozens of similar communications, and many hundreds of somewhat less in-

---

[31]The title for this essay was adapted from an article by Marc Bekoff entitled Compassionate conservation finally comes of age: Killing in the name of conservation doesn't work. The article was posted on August 20, 2010, at the Psychology Today website https://www.psychologytoday.com/blog/animal-emotions/201008/compassionate-conservation-finally-comes-age-killing-in-the-name.

sulting messages, were received after biologists working at my direction killed an adult female bear and sent
her two cubs to a zoo. This decision was made because
the bear killed a hiker and, with her cubs, subsequently
consumed a portion of the body. Her execution set off
a firestorm of protest, including inflammatory rhetoric, blatant misinformation, conspiracy theories, and
veiled threats.

Most people would be shocked to learn that animals are frequently killed in Yellowstone to further
long-term conservation goals, abide by cooperative
agreements with adjacent states, and protect residents
and visitors. Currently, about 300,000 nonnative lake
trout are captured in nets and killed each year in an
attempt to increase the survival and recruitment of native cutthroat trout in Yellowstone Lake. More than
1½ million lake trout have been killed since 2010 with
relatively little public protest. Also, hundreds to thousands of other nonnative fish are electrocuted or poi-
soned in rivers and streams each year to facilitate the
restoration and recovery of native fishes such as Arctic
grayling and cutthroat trout. In addition, more than
6,600 bison have been captured and shipped to meat
processing facilities since 1985 to protect the livestock
industry from the risk of brucellosis transmission—
even though the disease was originally introduced into
these wild animals from cattle. These shipments and
deaths result in outrage from the national and inter-
national public.

Other animals (e.g., bears) in the park are hazed from
campgrounds or developed areas and removed if they
become conditioned to human foods or target humans
for injury or death. Likewise, animals that become ag-

gressive with or chase people may be killed. For example, a wolf that chased a bicyclist was shot and a coyote that attacked a skier was killed. Also, some animals are inadvertently killed during captures for research activities, such as to fit radio collars and/or assess body condition and reproductive status. The ecological insights gained from these rigorously planned and well-executed field studies contribute greatly to the development of effective policies for the conservation, management, and restoration of wildlife. However, some people believe these activities should not occur if there is a chance an animal could be killed.

The remainder of this essay focuses on the infrequent killing of grizzly bears that attack people in the park, including the recent killing of a beloved bear dubbed Blaze by local photographers and wildlife watchers. Grizzly bears are one of the foremost attractions for visitors to the Yellowstone area. About 150 to 200 grizzly bears live primarily in the 2.2 million acres encompassed by Yellowstone National Park, yet conflicts with humans are infrequent because many bears have necessarily habituated to people. Thus, millions of visitors to the park watch and photograph grizzly bears without incident. However, tourism now exceeds 4 million visits per year, thereby increasing the risk of people surprising grizzly bears and provoking attacks.

Since 1970, there have been 53 attacks by grizzly bears on people in Yellowstone, and six people were killed.[32,33]

---

[32]Prior to 1970, one man (1916) and one woman (1942) were killed by grizzly bears in Yellowstone National Park.

[33]Gunther, K. 2015. Risk, frequency, and trends in grizzly bear attacks in Yellowstone National Park. Yellowstone Science 23:62-64.

Most of these attacks were likely defensive, with bears reacting to protect themselves, food, or cubs following surprise encounters with people. During most confrontations, the bears neutralized the perceived threat without killing the hiker(s) and then left the area with their cubs, if present. Only two attacks in 1984 (one fatality, one survivor) were thought to be predatory, with a bear silently approaching and attacking a person, presumably to obtain food. Thus, predatory attacks by grizzly bears on humans appear to be rare in Yellowstone.

Each of the six fatal attacks since 1970 was investigated to determine the circumstances, which were compiled in reports and reviewed by independent experts. If evidence indicated the attack was predatory, biologists attempted to capture and kill the bear. Conversely, if evidence indicated the attack was defensive, the bear was left in the wild unless it had subsequently fed on the body. No one knows if bears that consume human flesh are more likely to view people as food afterwards. However, bears are highly adaptable, intelligent animals that quickly learn and remember new foods. Thus, if after consuming human flesh, these bears viewed humans as relatively easy sources of fat and protein, they could kill another person and/or teach their cubs this behavior. As a result, managers have been cautious and removed them and their cubs from the wild.

Each decision regarding whether to kill a grizzly bear that attacked a person was made based on the gathered evidence and consideration of the perceived risk posed to visitors and staff. Considerations included whether the attack was defensive or predatory, the aggressiveness of the bear (e.g., charging distance), whether the attack was brief or prolonged, the bear's history of conflicts with humans, and whether the body was partially consumed and/or cached. These decisions were often difficult because the precise circumstances of most attacks were unknown, and no one knows the intent or motivation of a bear. Ultimately, four grizzly bears involved in the attacks on humans were killed by managers, primarily because each of these bears had subsequently fed on human flesh. These deaths upset many people who maintain the bears behaved naturally and were not at fault, had not had prior incidents with humans, and were not more likely to kill a person in the future. Also, many people argued hikers assumed the risk of an attack in grizzly country, and that several of the people killed were not following recommendations such as carrying bear spray, hiking in groups of three or more people, staying at least 100 yards away from bears, staying alert and on established trails, and making noise in areas with limited visibility.

Certainly, we should not blame a wild bear that responds to provocation by humans, whether intentional or not, by protecting itself, its food, or its cubs from a perceived threat. Also, the initial defensive attack and the later consumption and caching of the body are separate behavioral decisions, all of which are natural responses for a wild bear. However, the characterization by many people that these four bears were killed for retribution or to avoid lawsuits and bad press is wrong. A bear that consumes human flesh and caches a body to return later for further feeding is considering the human body as food, whether that was its initial intent for attacking or not. Perhaps, the bear and/or its cubs will never kill another person; perhaps they will. No one knows, so it's about how much risk we are willing to take. A bear left in the wilds of Yellowstone after it killed a man in a defensive response during 2011 was present at the fatality site of another man less than two months later, and probably fed on the body. Therefore, this bear was captured and killed, again with much public protest to leave the bear alone. However, park managers (including me) did not want to risk another person being killed. Leaving a bear in the wild that may now consider humans as food is an unnecessary risk to park visitors and overall public confidence and trust, especially considering the grizzly bear population currently consists of about 700 bears across 13 million acres in the Greater Yellowstone Ecosystem.[34, 35]

Most people don't come to the park to risk their lives. Naïve or not, they expect to see and photograph free-roaming animals that don't threaten their safety. When this expectation is not met, and people are terrified of wildlife rather than fascinated by them, it ultimately affects conservation because millions of them are taxpayers and voters. People that fondly recall their experiences in Yellowstone generally support the

[34]Bjornlie, D. D., D. J. Thompson, M. A. Haroldoson, C. C. Schwartz, K. A. Gunther, S. L. Cain, D. B. Tyers, K. L. Frey, and B. C. Aber. 2014. Methods to estimate distribution and range extent of grizzly bears in the Greater Yellowstone Ecosystem. Wildlife Society Bulletin 38:182-187.

[35]Haroldson, M. A., F. T. van Manen, and D. D. Bjornlie. 2015. Estimating number of females with cubs. Pages 11-20 in F. T. van Manen, M. A. Haroldson, and S. C. Soileau, editors. Yellowstone grizzly bear investigations: Annual report of the Interagency Grizzly Bear Study Team, 2014. U.S. Geological Survey, Bozeman, Montana.

park's conservation mission for the rest of their lives, even if they live in urbanized areas. However, support for the protection and further recovery of grizzly bears and their habitats may wane if bears regularly kill and feed on humans.[36] Also, there would almost certainly be more intrusive management and less tolerance for grizzly bears locally and regionally. In addition, more grizzly bears would likely be shot during chance encounters with fearful or nervous hunters and recreationists if animals that killed and consumed people were left in the wild.

Some self-named "compassionate conservationists" and others have asked park managers to refrain from killing animals in Yellowstone and, instead, emphasize the importance of individual lives.[37] Their suggested approach for developing a safe haven for wild animals involves increasing the education and awareness of people to alter confrontational behaviors and facilitate coexistence.[38] Admittedly, I was trained to manage for sustainable, viable populations of wildlife, rather than focusing on the health and well-being of each indi-

[36] Gunther, K. 2016. Why the Park Service killed a grizzly bear in Yellowstone. Backpacker http://www.backpacker.com/survival/bears/why-the-park-service-killed-a-grizzly-bear-in-yellowstone-2/

[37] Bekoff, M. 2015. Yellowstone kills Blaze, a bear who attacked off-trail hiker. Posted August 13, 2015, on the Psychology Today website at https://www.psychologytoday.com/blog/animal-emotions/201508/yellowstone-kills-blaze-bear-who-attacked-trail-hiker.

[38] Bekoff, M. 2015. Compassionate conservation meets Cecil the slain lion. Posted August 9, 2015, on the Psychology Today website at https://www.psychologytoday.com/blog/animal-emotions/201508/compassionate-conservation-meets-cecil-the-slain-lion.

vidual. Thus, I have some concerns and doubts about their premise and proposed approach. First, the apparent insinuation seems to be that Yellowstone biologists and managers are not compassionate because they occasionally kill bears that may pose a higher danger to humans. This criticism seems misplaced given that most of these people have dedicated their lives to preserving and restoring native populations of wildlife. Also, I can unequivocally state that I didn't come to Yellowstone to kill animals, and I'm not aware of anyone else who did—quite the opposite.

In addition, I struggle with statements by some compassionate conservationists about how increased acceptance and compassion for wildlife by humans will lead us to harmony and peaceful coexistence. This seems unlikely given that many humans and wild animals are not peaceful by nature, but instead compete and kill to survive. In turn, it seems unrealistic to expect that other animals will never be killed by humans to protect themselves or their interests. Moreover, I dispute the assertion by some people that Yellowstone is not fulfilling its mission because we have killed four adult grizzly bears and removed four cubs that consumed human flesh during the past 100 years. Yellowstone's record of recovering native species and ecosystem processes, including grizzly bears, is incomparable.

So, how do we move forward? Obviously, I need to learn more about compassionate conservation by discussing and debating this approach with willing proponents.[39, 40] Also, I have committed to meet with people opposed to the killing of grizzly bears in Yellowstone and have a rational discussion about management. We have a common goal to work toward, the sustained recovery of grizzly bears in the Yellowstone area and connectivity with public lands elsewhere, eventually including the Northern Continental Divide Ecosystem in northern Montana. Finally, I encourage people to carry bear spray and know how to use it. By itself, this action should reduce the number of deaths of humans and bears.[41]

[39] Wallach, A. D., M. Bekoff, M. P. Nelson, and D. Ramp. 2015. Promoting predators and compassionate conservation. Conservation Biology 29:1481-1484.

[40] Russell, J. C., H. P. Jones, D. P. Armstrong, F. Courchamp, P. J. Kappes, P. J. Seddon, S. Oppel, M. J. Rauzon, P. E. Cowan, G. Rocamora, P. Genovesi, E. Bonnaud, B. S. Keitt, N. D. Holmes, and B. R. Tershy. 2015. Importance of lethal control of invasive predators for island conservation. Conservation Biology DOI: 10.1111/cobi.12666.

[41] Gunther, K. 2016. Why the Park Service killed a grizzly bear in Yellowstone. Backpacker http://www.backpacker.com/survival/bears/why-the-park-service-killed-a-grizzly-bear-in-yellowstone-2/

SECTION TWO:
Management

Essay Five:

No Room at the Inn: The Curious Quandary of Yellowstone Bison

NPS Photo/Neal Herbert

The Senate recently passed legislation declaring bison our national mammal. Bison define our history because native peoples depended on them for their survival and spiritual well-being, while colonists exploited them to support their westward expansion across the country. In addition, lessons learned from the massive slaughter of bison during colonization, and subsequent efforts to recover them from the brink of extinction, helped define our national conservation and sporting ethics, which are still considered models for the rest of the world. Thus, I was shocked to arrive at Yellowstone and find bison are essentially constrained to the park and not allowed entry into adjacent states, unlike every other species of migratory wildlife. This bias against bison hinders their further recovery across the Greater Yellowstone Ecosystem, which inspired my reference to the Christmas story in the title for this essay. So, why are bison treated as second-class citizens?

Plains bison were nearly extirpated in the late 1800s as millions were shot by Euro-Americans colonizing western North America.[42] Only about two dozen bison remained in the Yellowstone area, all within the newly created (1872) national park. The dedicated protection and restoration of this population over the next century increased numbers to about 5,000 bison in 2015. This remarkable recovery represents one of the greatest conservation achievements in our nation's history. Today, Yellowstone bison are considered the only sustainable, wild population of plains bison due to their large numbers, high genetic diversity, and adaptive capabilities. Thousands of bison move across a vast, unfenced landscape, where they are exposed to competitors, predators, and challenging environmental conditions.[43]

Inside the 2.2 million acres encompassed by Yellowstone National Park, bison move freely and can access any and all of the habitats therein. However, the mountainous park does not contain substantial low-elevation habitats typically used by ungulates during winter when deep snow pack limits access to forage at higher elevations. As a result, some bison migrate to valleys outside of the park in search of forage, as do deer, elk, and pronghorn. Currently, almost all migrating bison move into the state of Montana; few migrate into Idaho or Wyoming. The number of migrating bison increases with bison density, winter severity, spring green-up

---

[42]Some of the information in this essay has been previously conveyed at the Yellowstone National Park website http://www.nps.gov/yell/learn/nature/bisonmgnt.htm.

[43]White, P. J., R. L. Wallen, D. E. Hallac, and J. A. Jerrett. 2015. Yellowstone bison: Conserving an American icon in modern society. Yellowstone Association, Yellowstone National Park, Mammoth, Wyoming.

of new vegetation, and learning.[44] However, federal and state employees confine bison migrating into Montana to relatively small areas adjacent to the park. Bison are not allowed to migrate farther or disperse to new areas because some of them are infected with the bacterial disease brucellosis, which was inadvertently introduced into bison and elk in the Yellowstone area when humans brought in domestic cattle for sustenance (milk, meat).[45] This disease could potentially be transmitted back to livestock, with subsequent adverse economic impacts to the industry. In addition, there are concerns about human safety and property damage, competition with livestock for grass, and a lack of funds for state management.[46] Interestingly, many elk that live in the Yellowstone area also carry brucellosis, have a large body size, and compete with cattle for grass; but their movements are not restricted.

Since the 1940s, Yellowstone bison have been misrepresented by livestock interests as the reservoir for brucellosis in the Greater Yellowstone Ecosystem and the primary risk for transmitting the disease to elk and cattle.[47] However, evidence indicates Yellowstone bison are not the villains. There have been no transmissions

[44]Geremia, C., P. J. White, J. A. Hoeting, R. L. Wallen, F. G. R. Watson, D. Blanton, and N. T. Hobbs. 2014. Integrating population- and individual-level information in a movement model of Yellowstone bison. Ecological Applications 24:346-362.

[45]Meagher, M., and M. E. Meyer. 1994. On the origin of brucellosis in bison of Yellowstone National Park: A review. Conservation Biology 8:645-653.

[46]Boyd, D. P. 2003. Conservation of North American bison: Status and recommendations. Thesis, University of Calgary, Alberta, Canada.

[47]Bidwell, D. 2010. Bison, boundaries, and brucellosis: Risk perception and political ecology at Yellowstone. Society and Natural Resources 23:14-30.

of brucellosis from wild bison to cattle, while elk have transmitted the disease to cattle more than 20 times since 2000.[48] Elk are more abundant and widespread, and the disease has spread throughout their populations in the Greater Yellowstone Ecosystem. Although elk are clearly the primary threat of brucellosis transmission to cattle, and the prevalence of the disease in elk is increasing, the livestock industry and associated regulatory agencies have maintained an unrelenting focus on bison.

Montana sued the National Park Service in 1995 when bison began migrating out of Yellowstone National Park onto state lands. The secretaries of Agriculture and the Interior and the Montana governor reached a settlement in 2000, creating the Interagency Bison Management Plan.[49] This plan calls for a target of around 3,000 bison, but the population has high reproductive and survival rates and numbers increase rapidly when conditions are good. As a result, the population has actually averaged more than 4,000 bison since 2000. Under the plan, managers limit bison abundance and distribution in Montana to lessen the

---

[48]Rhyan, J. C., P. Nol, C. Quance, A. Gertonson, J. Belfrage, L. Harris, K. Straka, and S. Robbe-Austerman. 2013. Transmission of brucellosis from elk to cattle and bison, Greater Yellowstone Area, USA, 2002-2012. Emerging Infectious Diseases 19:1992-1995.

[49]U.S. Department of the Interior, National Park Service, and U.S. Department of Agriculture, U.S. Forest Service and Animal and Plant Health Inspection Service. 2000. Bison management for the state of Montana and Yellowstone National Park. Final Environmental Impact Statement. Washington, D.C. Available at http://ibmp.info/.

probability they will mingle with cattle and transmit brucellosis. Capturing bison inside the park near the boundary and shipping them to meat processing facilities has been the principal method used to reduce abundance. The meat and hides from processed animals are distributed by Native American tribes to their members. Many people are outraged by these shipments, and park biologists and rangers don't want to do it either. As a result, managers are exploring alternatives such as quarantine to identify bison without brucellosis that could be sent alive to other places for conservation and cultural purposes.

In addition, managers are trying to reduce shipments by increasing public and tribal hunting of bison in Montana. In many winters, however, most bison do not migrate outside the park until March and April, when hunting seasons are closed due to females being late in pregnancy. Thus, Native American tribes with rights to hunt bison on some public lands outside the park want a larger bison population to stimulate earlier and larger migrations and, in turn, the number of bison available for harvest. When migrations are small or late, tribal hunters become frustrated and assert that treaty rights include hunting bison inside the park, a point that is encouraged by the state veterinarian and some others associated with the livestock community. However, Congress specifically prohibited hunting in Yellowstone National Park in 1894 (16 USC 26). Park managers and conservation supporters strongly oppose hunting in the park because it would affect the behavior of many different animals and drastically change the experiences of visitors.[50]

The Interagency Bison Management Plan has been effective at maintaining a viable population of bison in Yellowstone National Park and preventing the transmission of brucellosis from bison to cattle in Montana. The state has allowed more tolerance for bison in some areas to improve hunting and conservation. For example, the Governor of Montana recently issued a decision to allow limited numbers of Yellowstone bison to inhabit some public lands adjacent to the park throughout the year.[51] This decision could further the recovery of bison in the ecosystem if they are treated like other wildlife with minimal human intervention except when necessary to minimize conflicts with humans. However, state and local governments and many private landowners do not support unmanaged migrations of bison out of the park. If bison abundance was allowed to increase unregulated and they dispersed into cattle-occupied areas of Montana, it is likely those bison would be killed or removed by state employees and in organized hunts. Also, the state would likely rescind tolerance for bison, which would be a major setback for bison conservation.

Given these circumstances, the successful recovery of Yellowstone bison currently poses a conundrum, or catch-22, for managers at Yellowstone National Park.

---

[50]Wenk, D. N. 2012. Letter N1615(YELL) dated September 12, 2012 from the Superintendent of Yellowstone National Park to C. MacKay, Board of Livestock, and Dr. M. Zaluski, State Veterinaran, Helena, Montana. National Park Service, Yellowstone National Park, Mammoth, Wyoming.

[51]Bullock, S. 2015. Decision notice. Year-round habitat for Yellowstone bison. Environmental assessment, November 2015. Helena, Montana.

The park should maintain a population of at least 3,500 to 4,500 bison because this is the only ecologically and genetically viable population of plains bison in existence.[52, 53] However, the tribes and various conservation groups want even more bison to increase the number of bison available to hunt and further their recovery in the Greater Yellowstone Ecosystem. Therefore, even if park managers wanted to, they could not reduce bison numbers below this level without the inevitable lawsuits and petitions to list Yellowstone bison and/or plains bison as endangered pursuant to the Endangered Species Act. In fact, two such petitions were considered by the Fish and Wildlife Service during 2015. However, Montana currently has limited tolerance for bison near the park and Idaho and Wyoming have indicated they don't want wild bison at all. These states have traditionally exercised primary management authority over wildlife on National Forest System lands (16 USC 528) in the Yellowstone area, including the hazing of bison from these lands back into the park at times. In turn, the management of a population of 5,000 or more bison may not be sustainable in the long-term if they are prevented from migrating and required to forage year-round almost entirely in the park. Though no one knows for certain, it is likely that many if not most bison historically migrated to lower elevations during winter like the deer,

---

[52] Freese, C. H., K. E. Aune, D. P. Boyd, J. N. Derr, S. C. Forrest, C. C. Gates, P. J. P. Gogan, S. M. Grassel, N. D. Halbert, K. Kunkel, and K. H. Redford. 2007. Second chance for the plains bison. Biological Conservation 136:175-184.

[53] Hedrick, P. W. 2009. Conservation genetics and North American bison (Bison bison). Journal of Heredity 100:411-420.

elk, and pronghorn. If migrating bison are forced to remain within the park and relatively small nearby areas, numbers would be regulated by food availability within this area and bison would reach high densities before substantial starvation occurs. These high densities of bison could cause significant deterioration to other park resources such as vegetation, soils, geothermal features, and other ungulates as the bison population overshoots the park's capacity to provide adequate forage.[54]

So, how should we deal with this dilemma? Managers at Yellowstone National Park cannot conserve a viable population of bison on their own because when bison leave the park they are no longer under the agency's jurisdiction. Instead, their management becomes the prerogative of the respective states and the Forest Service on National Forest System lands. The risk of cattle getting brucellosis from bison is already low due primarily to management actions taken by federal and state management agencies to prevent mingling. It is hard to justify more intrusive control when bison are not the primary vector for transmission of brucellosis to livestock and, in fact, there has never been a detected transmission of brucellosis from bison to cattle.[55] Also, the infrequent outbreaks in cattle that have occurred due to elk have been quickly isolated and eradi-

[54]Plumb, G. E., P. J. White, M. B. Coughenour, and R. L. Wallen. 2009. Carrying capacity, migration, and dispersal in Yellowstone bison. Biological Conservation 142:2377-2387.

[55]Rhyan, J. C., P. Nol, C. Quance, A. Gertonson, A. Belfrage, L. Harris, K. Straka, and S. Robbe-Austerman. 2013. Transmission of brucellosis from elk to cattle and bison, Greater Yellowstone Area, USA, 2002-2012. Emerging Infectious Diseases 19:1992-1995.

cated due to diligent testing requirements for livestock producers.

The management of bison will likely always require ingenuity and hard work because they will eventually push up against the limits of tolerance in modern society as they attempt to migrate, disperse, and/or expand their range. In other words, they will not stay put. Also, maintaining a post-winter population of 3,500 to 4,500 bison means that about 700 to 1,000 calves will be born each spring.[56] It is unlikely additional tolerance for bison on public lands in the states surrounding the park will keep pace with this rapid population growth. Given the extremely high survival of these calves and adults, it would be necessary to remove a similar number of bison each winter to maintain a relatively stable population size. As of 2015, public and tribal hunting had removed a maximum of 322 bison during a single winter. Even if this total can be increased substantially in coming years, it is likely that some level of culling will be necessary into the foreseeable future. Hunting could be facilitated by moving the existing capture facilities from the boundary of the park to the furthest extent of the tolerance area so bison can distribute more widely and be hunted more effectively before captures are attempted.

For long-term conservation and further recovery, bison need similar access to habitat and tolerance that other wildlife species such as elk are given in the Yellowstone area, including year-round access to many

---

[56]Geremia, C., R. Wallen, and P. J. White. 2015. Population dynamics and adaptive management of Yellowstone bison. National Park Service, Yellowstone National Park, Mammoth, Wyoming. Available at <ibmp.info>.

National Forest System and other public lands. Hunting practices also need to be adjusted to eliminate firing lines at the park boundary and allow bison to distribute across this larger landscape. This new paradigm would accommodate more bison and allow them to move more freely among suitable public lands in the Greater Yellowstone Ecosystem. I am not suggesting bison be restored to the entirety of their historic range, which is infeasible in modern society for a multitude of reasons. However, bison could be restored to a much larger portion of public lands in the Yellowstone area. Tourism is a major driver of the regional economy and public opinion is shifting toward more tolerance for bison outside preserves. Also, this increased distribution will not substantially increase the risk of brucellosis transmission from bison to livestock, provided landowners and managers take the same brucellosis prevention measures recommended in areas with wild elk populations chronically infected with brucellosis (i.e., hazing of elk to maintain separation with cattle; fencing of haystacks; targeted hunts to disperse elk, etc.).[57] Allowing bison to occupy more public lands would create new opportunities for hunting, bolster tourism, and enhance conservation.

Finally, only a few other unfenced, wide-ranging populations of plains bison exist in the United States today (e.g., Book Cliffs and Henry Mountains, UT; Grand Teton, WY; Wrangell-St. Elias, AK), all of which have less than 1,000 bison. Most other conservation herds of bison on public lands also have low population sizes,

---

[57] Montana Fish, Wildlife & Parks. 2014. Elk management in areas with brucellosis. 2015 proposed work plan. August 7, 2014 Fish and Wildlife (FW) Commission meeting, Helena, Montana.

along with limited distributions, protection from natural selection factors like large predators, and skewed sex and age ratios maintained to ease management.[58] Thus, additional wild, wide-ranging populations subject to the forces of natural selection need to be augmented or established at other sites to preserve the species. I realize this is a difficult task in modern society, but it should be feasible in some suitable areas and would reduce the reliance on Yellowstone and a few other populations to preserve the species in the wild.

[58]McDonald, J. L. 2001. Essay: Bison restoration in the Great Plains and the challenge of their management. Great Plains Research 11:103-121.

Essay Six:

When the Mission Runs into Tradition: The Influence of Sport Fishing

NPS Photo./Neal Herbert

When I first came to Yellowstone, I worked almost exclusively on birds and mammals. I had grown up fishing for bass, pickerel, and walleye in upstate New York, but had no experience in managing fisheries and no great passion to cast a line for the cherished trout of the Yellowstone area. In fact, the local guides and visiting anglers were less than charitable when I waded into the waters with my old sneakers, cut-off shorts, tank top and, God forgive me, bait fishing pole. Even with this lack of sophistication, however, it didn't take long for me to notice a dichotomy in how fish were managed inside Yellowstone National Park compared to other wildlife. Angling is allowed in the park, whereas hunting is not. Also, nonnative fish species were not only tolerated, but promoted in many waters of the park, again in contrast to other nonnative wildlife. Why do these differences in management and protection exist?

The answer to this question apparently stems from the Yellowstone National Park Protection Act (also known as the original Lacey Act) of 1894, in which

Congress specifically prohibited hunting in the park to protect remnant populations of birds and mammals from poaching (16 USC 26). This Act also prohibited residents and visitors from removing fish from waters in the park using drugs, explosives, nets, seines, or traps. However, the Act allowed angling in the park by hook and line under rules and regulations promulgated by the Secretary of the Interior (16 USC 26). As a result, angling continued and the stocking of native and nonnative fish into park waters was promoted to attract more visitors.[59]

Almost half of the lakes and waterways in Yellowstone had no fish in them when the park was established (1872) because waterfalls and other barriers prevented fish movements from downstream.[60] With good intentions, park managers began stocking many of these waters with nonnative trout to increase the variety of angling opportunities for visitors. They released nonnative trout species that originated from ecosystems thousands of miles away and even from another continent, such as brown trout from Europe, rainbow trout from Alaska and the west coast, eastern brook trout, and lake trout from the Great Lakes. These nonnative fish were also released into waters inhabited by native fish. Over time, more than 310 million fish were stocked in many lakes and watercourses, and Yellow-

[59]Franke, M. A. 1996. A grand experiment. 100 years of fisheries management in Yellowstone: Part I. Yellowstone Science 4:2-12.

[60]Evermann, B. W. 1892. Report on the establishment of fish-cultural stations in the Rocky Mountain Region and Gulf States. Bulletin of the U.S. Fish Commission for 1891. Government Publishing Office, Washington, D.C.

stone became a premier destination for sports fishing.[61]

By the 1950s, park managers realized that repeatedly stocking fish to support increasing numbers of anglers was not sustainable and stopped the practice. Harvest rates of cutthroat trout in Yellowstone Lake decreased substantially as angler numbers kept increasing. Also, the collection of tens of millions of eggs each year to raise cutthroat trout for stocking had collapsed spawning migrations in some streams. In addition, managers implemented new regulations during the late 1960s and 1970s to protect native fishes and promote catch-and-release instead of harvest.[62] Native fish populations began to recover under this new approach, but by the mid-1990s threats were increasing from competition, hybridization, and predation by nonnative fish. Also, predatory lake trout were illegally introduced into Yellowstone Lake, which supported the world's largest population of Yellowstone cutthroat trout.[63] To understand and address these threats, fisheries biologists increased research to obtain key information and began pilot projects to suppress nonnative fish in some areas and restore native species. In addition, biologists developed an innovative and comprehensive plan to

[61]Varley, J. D., and P. Schullery. 1998. Yellowstone fishes: Ecology, history, and angling in the park. Stackpole Books, Mechanicsburg, Pennsylvania.

[62]Gresswell, R. E., and J. D. Varley. 1988. Effects of a century of human influence on the cutthroat trout of Yellowstone Lake. Pages 45-52 in R. E. Gresswell, editor. Status and management of Interior stocks of cutthroat trout. American Fisheries Society, Bethesda, Maryland.

[63]Kaeding, L. R., G. D. Boltz, and D. G. Carty. 1996. Lake trout discovered in Yellowstone Lake threaten native cutthroat trout. Fisheries 21:16-20.

restore the ecological role of native fishes, while sustaining angling and fish watching experiences. Historically, thousands of people each year watched spawning cutthroat trout at Fishing Bridge and moving through LeHardy Rapids.

The Native Fish Conservation Plan was released in 2010 and included actions to recover numbers of cutthroat trout in Yellowstone Lake by suppressing lake trout, maintain and restore Yellowstone cutthroat trout in some river systems in northeastern Yellowstone, and restore Arctic grayling and westslope cutthroat trout in watersheds in northwestern Yellowstone.[64] Managers also revised angling regulations to require the killing of captured nonnative fish in some areas and ensure the release of native fish. Thereafter, efforts to suppress lake trout were intensified, with the killing of about 1½ million lake trout in Yellowstone Lake and, in turn, increased recruitment of juvenile cutthroat trout. Arctic grayling and westslope cutthroat trout were restored to the Grayling Creek drainage, and westslope cutthroat trout were restored to the Specimen Creek drainage, in northwest Yellowstone after removing nonnative trout. Likewise, Yellowstone cutthroat trout were returned to the Elk, Yanceys, and Lost Creek stream complex and Soda Butte Creek in northeast Yellowstone after removing nonnative brook trout through the application

---

[64]Koel, T. M., J. L. Arnold, P. E. Bigelow, and M. E. Ruhl. 2010. Native fish conservation plan. Environmental assessment, December 16, 2010. National Park Service, Department of the Interior, Yellowstone National Park, Wyoming. http://parkplanning.nps.gov/document.cfm?parkID=111&projectID=30504&documentID=37967.

of piscicides, or fish poisons. In addition, actions were taken to suppress and prevent the expansion of non-native fish on portions of the Lamar River and Slough Creek to protect native cutthroat trout.[65]

These restoration efforts were strongly supported by ecologists, federal and state agencies, and much of the sports fishing community. However, there was widespread concern among some fly fishermen about regulations requiring the killing of nonnative fish in some areas, contrary to the catch-and-release ethic that has been championed in Yellowstone and throughout much of the nation for decades. Also, a small, outspoken contingent of sports fishermen, fly shop owners, and guides strongly opposed the suppression of nonnative fishes, which they argued had coexisted with natives for many decades without adverse effects. In addition, there was concern about the use of the chemical rotenone to remove nonnative fishes because it can kill amphibians and aquatic invertebrates. Several people mistakenly purported rotenone has life-threatening effects to other wildlife and humans. Thus began the inevitable conflict between the park's mission to preserve and restore native species and the long-standing tradition of sports fishing for favored species of nonnative

---

[65]Koel, T. M., J. L. Arnold, P. E. Bigelow, C. R. Detjens, P. D. Doepke, B. D. Ertel, and M. E. Ruhl. 2015. Native Fish Conservation Program, Yellowstone Fisheries & Aquatic Sciences 2012-2014. Report YCR-2015-01, National Park Service, Yellowstone National Park, Mammoth, Wyoming.

trout.[66] Furthermore, there was opposition from some park staff regarding the construction of nonnative fish barriers and the use of motorized boats in wilderness areas. Also, fish restoration projects sometimes conflicted with other preservation efforts, for example when loons were caught in nets set to remove the nonnative lake trout.

Some park biologists and managers were a bit surprised by this opposition. Obviously, we did not appreciate the attitudes and desires of traditional anglers, primarily fly fishermen, in Yellowstone. To rectify the situation, we reached out to Trout Unlimited, Greater Yellowstone Coalition, National Parks Conservation Association, Yellowstone Park Foundation, and other stakeholder groups to discuss ongoing and proposed projects and to partner on funding, messaging, and research related to ongoing and future restoration efforts. Leaders and scientists from these organizations were instrumental in developing and disseminating articles, editorials, educational materials, and letters to counter misinformation and misperceptions about the restoration activities.[67] Also, the park's fisheries staff

---

[66] This dilemma and the actions taken and lessons learned described in this essay were previously published in an abstract by D. E. Hallac, P. Bigelow, T. Koel, and P. J. White in 2014 entitled Noble versus Native—The Societal Challenges of Restoring Native Trout in Yellowstone National Park (Pages 101-102 in R. F. Carline and C. LoSapio, editors. Wild Trout XI: Looking back and moving forward. Wild Trout Symposium, West Yellowstone, Montana).

[67] For example, see the document Science Supporting Management of Yellowstone Lake Fisheries: Responses to Frequently Asked Questions by Trout Unlimited at http://wyomingtu.org/wp-content/uploads/2014/03/Science-Supporting-Management-of-Yellowstone-Lake-Fisheries.pdf.

reinstituted public meetings in gateway communities around the park each spring to discuss fisheries management and angling regulations. In addition, park biologists and managers met with fly shop owners and guides to discuss apprehensions and concerns, as well as with private landowners worried about the use of rotenone. Staff provided landowners with the results of water sampling conducted in streams and wells before and after treatments to reduce fears regarding poisoning.

These actions were effective at disseminating information, countering falsehoods, and better involving stakeholders in the evaluation and decision-making processes. However, we had little success changing the opinions of fishermen, fly shop owners, and guides who prefer angling for nonnative fish and, not surprisingly, prioritize their enjoyment and economic well-being over the park's mission to preserve native fish. Most of these people were either not familiar with the park's mission and/or didn't support it when it conflicted with their favored fishing locations or practices. Similarly, several park staff had issues with the extent of short-term impacts in wilderness areas necessary to complete the mission, even though the intent of restoration was for long-term gain. Lastly, we had little to no success in convincing fly fishermen to embrace the killing of nonnative fish due to their catch-and-release creed, including several park managers.

One issue that confuses or frustrates some fly fishers is the penchant of park biologists for genetic purity when existing native trout are only slightly hybridized from inbreeding with nonnative species and, as a result, possess a very small portion of nonnative genes.

In these situations, removing nonnative or hybridized fish may be more palatable to stakeholders when existing native or slightly hybridized fish are nonlethally removed prior to chemical treatment, held in a protected area during treatments, and then returned afterwards. Biologists from Yellowstone National Park and collaborators can then augment these fish with genetically pure fish thereafter. Another issue that irritates wilderness proponents is the intrusive nature of restoration projects in wilderness and the length of time these disturbances may need to be sustained to achieve restoration. Installing barriers and removing nonnative fish often requires several trips with helicopters to haul in construction materials and maintain crews in certain remote areas for many weeks or months. There needs to be continuing dialogue on the wilderness issue to find common ground since both restoration and wilderness proponents are trying to preserve naturalness and wildness. For long-term or large-scale projects such as the suppression of lake trout in Yellowstone Lake or restoration actions in the Lamar and upper Gibbon watersheds it is imperative to continue independent scientific review of activities and their effectiveness. This review should be enhanced by the perspectives of conservation groups (e.g., Trout Unlimited) and the fly fishing community (owners/ guides). In turn, there needs to be more recognition by some stakeholders of the park's mission, with fishing in Yellowstone being more about the unique experience in a wilderness setting than maximizing landing rates.

The Native Fish Conservation Plan identified several rivers, including the Firehole and Madison rivers, and sections of the Gibbon River, where native communities cannot be feasibly restored and, as a result, they will remain nonnative trout fisheries. Currently, restoration activities in other areas involve less than 10 percent of the park's waterways, with traditional sport fishing continuing in the remainder. Personally, I doubt any amount of dialogue or cajoling will convince many fly fishers to implement the must-kill regulations for nonnative fishes in certain areas. Likewise, I don't believe that spending more time with fly shop owners and guides will produce many converts. However, I recommend park managers continue this outreach and invite these stakeholders into restoration partnerships to increase their involvement and input into the evaluation and decision-making process. These partnerships would facilitate the transfer of information and ideas, as well as provide advanced notice of forthcoming restoration activities. Over time, these partnerships should increase respect and trust among the various parties, and lessen the amount of frustration and misinformation. In addition, there needs to be more emphasis on disseminating the remarkable conservation success stories and new angling opportunities created by these restoration efforts. This outreach should highlight the contributions of the partners, their various reasons for supporting the projects, and pride in preserving and restoring native fishes as well as tradition.

In all honesty, if I could turn back the clock I would probably support the prohibition of fishing in Yellowstone, just like hunting. I believe the value of Yellowstone as a preserve where wildlife species live in a relatively undisturbed state with education, learning, observation, and the attainment of scientific knowledge by people outweighs the immediate benefits and enjoyment from angling and hunting. Besides, wildlife migrating and dispersing from the park already support productive recreational and sporting activities in outlying areas. However, the Yellowstone National Park Protection Act established a precedent for angling in the park, and this tradition has continued in various forms throughout the park's history. In fact, this tradition has led to effective paradigms for fisheries management that are currently promoted throughout the world. Therefore, I accept the reality of catch-and-release angling in the park and appreciate the partnerships with various angling organizations that have greatly contributed to fisheries management and native species restoration activities therein. However, I believe a primary focus of future fisheries management should be to promote and recover native species in support of the park's mission, which would include the removal of nonnative fishes where feasible. Native fish cannot fulfill their ecological role in the park if their populations are decimated, hybridized, or isolated.[68] The tradition of angling for nonnative species could continue in areas where native restoration is not achievable due to biological or social reasons.

---

[68]Varley, J. D., and P. Schullery. 1998. Yellowstone fishes: Ecology, history, and angling in the park. Stackpole Books, Mechanicsburg, Pennsylvania.

NPS Photo/Neal Herbert

Essay Seven:

King of the Mountain: Grizzly Bear Recovery and Their Uncertain Future

When I first came to Yellowstone, more than anything else I wanted to see a grizzly bear. To me, they are emblematic of wild places, moving fearlessly across the landscape, dominant over humans and other animals. They remind me of our ancestors who lived in a natural world filled with dangers and difficulties rarely experienced by most people today. Also, their size, speed, and strength evoke mystical reverence, while their explosive aggression in defense of food or young generates terror. These magnificent animals were isolated and threatened with extinction in the Yellowstone area by the mid-1970s due to conflicts with humans and habitat loss. However, decades of protection and dedicated, innovative management reversed this trend and the bears made a remarkable recovery. Given this tremendous success, why is there still controversy about their management?

Early in the park's history, many black bears and grizzly bears were troublesome garbage raiders, whose begging along roads, feeding at dumps, and forays through campgrounds and residential areas attracted people from around the globe.[69] Unfortunately, bears expecting handouts become aggressive when not rewarded and, over time, increasing human injuries led to changes in management that ended the reliance of bears on human foods and, in the long term, restored them as wildlife.[70] In the short term, however, these changes led to many food-conditioned bears being captured and killed or sent to zoos. By the mid-1970s, fewer than 220 grizzly bears remained in and near Yellowstone and numbers were decreasing, primarily due to conflicts with humans.[71] As a result, Yellowstone grizzly bears were protected as threatened under the Endangered Species Act and managers began implementing actions to restore a viable population. This recovery took many decades, but eventually bears doubled their occupied range and more than tripled in abundance. Today, there are about 700 grizzly bears inhabiting more than 13 million acres in the Greater Yellowstone Ecosystem.[72,73]

---

[69] Essay Seven was written with Kerry Gunther, Bear Management Biologist, Yellowstone National Park.

[70] Schullery, P. 1992. The bears of Yellowstone. High Plains Publishing Company, Worland, Wyoming.

[71] Meagher, M. 2008. Bears in transition, 1959-1970s. Yellowstone Science 16:5-12.

[72] Bjornlie, D. D., D. J. Thompson, M. A. Haroldoson, C. C. Schwartz, K. A. Gunther, S. L. Cain, D. B. Tyers, K. L. Frey, and B. C. Aber. 2014. Methods to estimate distribution and range extent of grizzly bears in the Greater Yellowstone Ecosystem. Wildlife Society Bulletin 38:182-187

[73] Haroldson, M. A., F. T. van Manen, and D. D. Bjornlie. 2015. Estimating number of females with cubs. Pages 11-20 in F. T. van Manen, M. A. Haroldson, and S. C. Soileau, editors. Yellowstone grizzly bear investigations: Annual report of the Interagen-

Given that the population is biologically recovered and sustainable, the Fish and Wildlife Service is considering whether to delist, or remove, grizzly bears in the Yellowstone area from protection under the Endangered Species Act. If the Fish and Wildlife Service takes this action, bears would be managed by the Forest Service, National Park Service, Bureau of Land Management, Native American Tribes, and the states of Idaho, Montana, and Wyoming on their respective lands pursuant to an agreed-upon Conservation Strategy to ensure the continued viability of the population. Humans continue to be the primary cause of death for grizzly bears due to management removals (depredation of livestock, food conditioning), defense of life or inadvertent kills by hunters, and vehicle strikes, which results from increasing habitat encroachment from dwellings, recreation, and visitation. As a result, lessening conflicts with people will be essential for maintaining and continuing the recovery of grizzly bears. Also, there are emerging challenges that could complicate the conservation of grizzly bears, including a warming climate, as well as substantial consternation from bear advocates and some Native American tribes about the possible initiation of trophy harvests after delisting.

Continued climate warming could adversely affect several key food resources for grizzly bears that have already decreased during recent decades, including elk, whitebark pine seeds, and Yellowstone cutthroat trout. Some scientists believe the dietary resiliency of grizzly bears will enable them to adapt to these chang-

cy Grizzly Bear Study Team, 2014. U.S. Geological Survey, Bozeman, Montana.

ing conditions, while others contend a substantial decrease in the abundance and distribution of grizzly bears is inevitable due to poorer nutrition and higher mortality risk. Recent studies by members of the Interagency Grizzly Bear Study Team indicate grizzly bears eat a diversity of plant and animal species, which provides some flexibility to respond to changing food resources.[74] As a result, bears have compensated for recent decreases in cutthroat trout and whitebark pine seeds by shifting to other nutritious foods without a loss in body condition.[75] However, there is still uncertainty regarding the future extent of climate changes, the magnitude of its effects on grizzly bears, and the resilience of bears to adapt to these changes. As a result, monitoring and research will continue to provide early detection of any adverse changes and trends.

Grizzly bears are one of the premier wildlife attractions for visitors to Yellowstone, which contributes to the public's enjoyment and sense of pride in our conservation heritage, as well as infusing millions of dollars into the regional economy. Thus, one of the most controversial issues surrounding delisting is state-managed trophy hunts near the boundaries of Grand Teton and Yellowstone National Parks, and the potential harvest of beloved, recognizable grizzly bears that primarily live in these parks, but at times venture

---

[74]Gunther, K. A., R. R. Shoemaker, K. L. Frey, M. A. Haroldson, S. L. Cain, F. T. van Manen, and J. K. Fortin. 2014. Dietary breadth of grizzly bears in the Greater Yellowstone Ecosystem. Ursus 25:60-72.

[75]van Manen, F. T., M. A. Haroldson, D. D. Bjornlie, M. R. Ebinger, D. J. Thompson, C. M. Costello, and G. C. White. 2016. Density dependence, whitebark pine, and vital rates of grizzly bears. Journal of Wildlife Management 80:300-313.

beyond. Many grizzly bears in these parks have necessarily habituated to the presence of millions of non-threatening visitors each year, which has likely made them somewhat naïve and susceptible to hunting when they leave the parks. Therefore, the harvest of some of these habituated bears will be highly contentious.

Biologists and managers at Yellowstone have recommended to the Fish and Wildlife Service and the surrounding states that any future harvests be conducted in a way that (1) respects the mission of the National Park Service, (2) protects regional economic benefits and the enjoyment of bear watching, (3) reduces the risks associated with wounded bears entering the parks, and (4) limits the likelihood that well-known bears will be harvested. If and when grizzly bears are delisted, Yellowstone managers have requested to be included in annual meetings regarding the allocation of harvest mortality by the states and asked the states to focus the majority of harvests away from park boundaries, such as in areas where human-bear conflicts are prevalent.

In addition, Yellowstone managers requested the annual harvest allocation be based on the number of grizzly bears available to be hunted outside park units—not the entire population. About 21% of the Demographic Monitoring Area for Yellowstone grizzly bears is designated as National Park Service units, where the hunting of bears will not be allowed. Basing the mortality allotment on total abundance (100% of the area), when only a portion of the population (~79%) is available for harvest outside national parks, would concentrate harvest on this population segment and could reduce dispersal and connectivity. Basing the harvest on the number of grizzly bears available to be hunted outside park units would help lessen the adverse effects of habitat loss from private developments in the area, dampen the effects of high and unpredictable human-caused mortality on multiple use and private lands (i.e., provide a refuge effect), and offset the effects of unreported, harvest-related, wounding loss on the grizzly bear population.

Currently, the grizzly bear population in the Yellowstone area appears to be varying near the capacity of the environment to support bears, with the growth rate of the population slowing over the past decade as bear density increased. [76] Efforts to reduce conflicts with people and preserve habitat for dispersal and, eventually, connectivity with other populations will be essential for further restoration. The recommendations provided herein should allow for periods of increasing and decreasing abundance within the ecosystem, as well as the potential for grizzly bears to pioneer additional areas and eventually connect with other bears in the Northern Continental Divide Ecosystem in northern Montana and Canada.

The successful conservation and recovery of grizzly bears in the Greater Yellowstone Ecosystem highlight how the expectations and perceptions of society about wildlife have changed over time, including how people want to interact with and enjoy wildlife (e.g., viewing bears from bleachers at garbage dumps versus seeing bears in natural settings going about their lives). Wild-

---

[76] van Manen, F. T., M. A. Haroldson, D. D. Bjornlie, M. R. Ebinger, D. J. Thompson, C. M. Costello, and G. C. White. 2016. Density dependence, whitebark pine, and vital rates of grizzly bears. Journal of Wildlife Management 80:300-313.

life science contributed to this progression by enhancing our ecological understanding of bears and adapting management as this new understanding accrued and influenced public attitudes and managers to adopt new and more enlightened policies.

NPS Photo/Daniel Stahler

Essay Eight:

"I'll Huff, and I'll Puff:" The Rhetoric and Reality of Wolf Restoration

NPS Photo/Daniel Stahler

One of my first tasks after arriving at Yellowstone National Park was to analyze the effects of reintroduced wolves on elk, their primary prey. These effects continue to be heatedly argued among laypersons, politicians, and scientists, with claims ranging from wolves annihilating elk throughout the region to their restoration healing the entire ecosystem. Though preeminent wolf biologist Dave Mech cautioned that "wolves are neither saints nor sinners,"[77] the fervor and contrast in how many people view wolves continues to range from complete support and protection to total intolerance (i.e., "the only good wolf is a dead wolf"). Thus, the threats by the wolf in The Three Little Pigs nursery rhyme[78] typify the bluster and rhetoric from both advocates and critics. The truth is that wolves are neither

---

[77]Mech, L. D. 2012. Is science in danger of sanctifying the wolf? Biological Conservation 150:143-149.

[78] Halliwell, J. O. 1886. The nursery rhymes of England. Frederick Warne and Company, London and New York.

vindictive killers nor thoughtful redeemers—they are merely doing what comes natural to them—making a living and interacting with other animals in the area.

Wolves were eliminated from the Yellowstone area by the 1940s by settlers, soldiers, and others who feared them and loathed them for killing livestock and wild-life species needed for sustenance such as elk. The loss of wolves and drastic reductions in other large pred-ators such as bears and cougars led to changes in the ecosystem, as elk numbers proliferated and browsing effects on preferred forage species altered vegetation communities. By the 1930s, managers believed that overgrazing was having a deleterious effect on vegeta-tion in the park. As a result, they removed more than 70,000 elk from the northern Yellowstone area over the next 40 years through culling in the park and hunting in Montana, which gradually decreased elk counts to fewer than 4,000 by 1968.[79] Thereafter, culling ceased inside the park and counts of northern Yellowstone elk rapidly increased to about 19,000 in two decades.

At that time, the park's high mountains and plateaus provided summer range for tens of thousands of elk from about eight different populations. Most of these elk spent winter in lower elevation valleys outside the park, where snow depths were much lower, but migrat-ed into the park during spring to access high-quality forage growing in the park's productive and extensive grasslands. Visitors traveling through the area had an exceptional opportunity to observe hundreds to thou-sands of elk moving across the vast landscape, com-peting for breeding opportunities during the rut, and

---

[79]Houston, D. B. 1982. The northern Yellowstone elk herd. Macmillan, New York, New York.

nursing and protecting their newborn calves. Also, sportsmen and women flocked to the region to experience the highly successful and magnificent wilderness hunts. As a result, elk in the Yellowstone area were treasured, and tourism and hunting had a large influence on the regional economy.

In addition, migratory elk played a prevalent role in the Yellowstone ecosystem by transferring nutrients across the landscape, converting grass to animal tissue, competing with other grazers, and providing sustenance for predators, scavengers, and decomposers. Low to moderate grazing increased the availability of light, moisture, nitrogen, and other nutrients to grasses, which stimulated the growth of new tissue and the overall productivity of the grasslands.[80] Elk also represented a key food source for species ranging from bears to magpies to beetles and bacteria in the soil. In the end, their carcasses contributed to nutrient surges that greatly enhanced the productivity of nearby plants.[81]

However, growing numbers of elk eventually led to bitter debates about them irreversibly damaging the landscape. Intensive grazing by large numbers of elk can decrease productivity by removing too much vegetation, compacting soils, and reducing the diversity of

[80]Frank, D. A., R. L. Wallen, and P. J. White. 2013. Assessing the effects of climate change and wolf restoration on grassland processes. Pages 195-205 in P. J. White, R. A. Garrott, and G. E. Plumb, editors. Yellowstone's wildlife in transition. Harvard University Press, Cambridge, Massachusetts.

[81]Garrott, R. A., D. A. Stahler, and P. J. White. 2013. Competition and symbiosis. The indirect effects of predation. Pages 94-108 in P. J. White, R. A. Garrott, and G. E. Plumb, editors. Yellowstone's wildlife in transition. Harvard University Press, Cambridge, Massachusetts.

plants. Over time, intensive browsing and grazing can transform vegetation communities by shifting their composition and structure. An independent investigation by the National Research Council concluded the grasslands in Yellowstone were not overgrazed,[82] but the debate continued until wolves were reintroduced to the ecosystem from 1995 to 1997. Wolf numbers and distribution increased rapidly, with elk comprising the majority of their kills. As elk numbers decreased over the next two decades, concerns changed from there being too many elk to there being too few elk to support hunting opportunities in surrounding states. As numbers of counted elk in northern Yellowstone decreased to less than 5,000 by 2011, there was outrage from sportsmen and local business owners and outfitters. Elk permits were reduced several times by state wildlife agencies trying to promote higher recruitment and survival, and these reductions provoked angry calls for liberal harvests of predators within and near the park.

There is no doubt that wolf abundance and distribution increased far more rapidly than biologists predicted, from 21 released in Yellowstone in 1995 to 174 within the park, and hundreds more in surrounding states, in less than 10 years. Surprisingly, wolf kill rates of elk remained high, and the population continued to grow rapidly, even after a 50% decrease in the number of elk. Also, the numbers of elk spending winter in Yellowstone eventually decreased (75%) more than the

---

[82]National Research Council. 2002. Ecological dynamics on Yellowstone's northern range. National Academies Press, Washington, D.C.

5-30% forecast by most biologists before wolf reintroduction.[83] These decreases were especially evident in high-elevation areas with high densities of predators, deep snows, and terrain characteristics that made elk more vulnerable to predation.[84, 85]

However, wolves were not solely responsible for the decrease in elk numbers, as is widely advocated. In fact, wolves were not even the predominant mortality source during the first 5 to 7 years after reintroduction when their numbers were relatively low, elk numbers were very high and, as a result, wolf predation by itself had little effect on the population dynamics of elk.[86] Surprisingly, biologists found that bear predation on newborn elk calves was extensive and much higher than previously thought, which had a substantial effect on recruitment in some areas.[87] Moreover, harvests of elk remained high for almost a decade following wolf restoration, including late season hunts of primarily

[83]White, P. J., and R. A. Garrott. 2005. Yellowstone's ungulates after wolves—Expectations, realizations, and predictions. Biological Conservation 125:141-152.

[84]Hamlin, K. L., R. A. Garrott, P. J. White, and J. A. Cunningham. 2009. Contrasting wolf- ungulate interactions in the Greater Yellowstone Ecosystem. Pages 541–577 in R. A. Garrott, P. J. White, and F. G. R. Watson, editors. The ecology of large mammals in central Yellowstone: Sixteen years of integrated field studies. Elsevier, San Diego, California.

[85] Dunkley, S. L. 2011. Good animals in bad places: Evaluating landscape attributes associated with elk vulnerability to wolf predation. Thesis, Montana State University, Bozeman, Montana.

[86]Smith, D. W., R. O. Peterson, and D. B. Houston. 2003. Yellowstone after wolves. BioScience 53:330-340.

[87]Barber- Meyer, S. M., L. D. Mech, and P. J. White. 2008. Elk calf survival and mortality following wolf restoration to Yellowstone National Park. Wildlife Monographs 169.

prime-aged female elk with high reproductive value. Also, persistent drought conditions from 1999 through 2007 may have reduced the nutrition of elk and contributed to lower maternal condition, pregnancy rates, calf survival, and recruitment.[88] By the early to mid-2000s, however, wolves had become the predominant factor influencing the dynamics of elk and, in combination with bears and cougars, facilitated and sustained low recruitment and overall numbers of elk.[89]

Another unexpected twist after wolf reintroduction was that Yellowstone quickly became one of the premier places in the world to watch wild wolves due to their visibility from roads. Surprisingly, many wolves became quite tolerant of people and moved about the landscape apparently unaffected by their presence. As a result, wolves are now one of the primary reasons people come to the park and a great source of enjoyment to millions of people around the world. By themselves, wolf watchers contribute tens of millions of dollars each year to local economies.[90] Many of the wolves in Yellowstone are well-known because they are

[88]Cook, J. G., B. K. Johnson, R. C. Cook, R. A. Riggs, T. Delcurto, L. D. Bryant, and L. L. Irwin. 2004. Effects of summer-autumn nutrition and parturition date on reproduction and survival of elk. Wildlife Monographs 155.

[89]Proffitt, K. M., J. A. Cunningham, K. L. Hamlin, and R. A. Garrott. 2014. Bottom-up and top-down influences on pregnancy rates and recruitment of northern Yellowstone elk. Journal of Wildlife Management 78:1383-1393.

[90]Duffield, J. W., C. J. Neher, and D. A. Patterson. 2008. Wolf recovery in Yellowstone: Park visitor attitudes, expenditures and economic impacts. Yellowstone Science 16:21-25.

individually identifiable and quite visible to wildlife watchers. Consequently, many people are passionate about observing and protecting wolves in Yellowstone.

One fiery issue for these advocates is the hunting and trapping of wolves outside, but near, Yellowstone National Park. Many people are outraged when beloved, recognizable wolves that primarily live in the park move outside and are legally harvested in surrounding states. The park's primary objective is to maintain naturally functioning wolves by minimizing human intervention within the park, which is still achievable with a modest harvest outside the park, similar to other wildlife species. Wolves living in Yellowstone National Park are part of a larger population that includes much of the northern Rocky Mountains. At the end of 2014, there were 510 wolves in at least 73 packs in the Greater Yellowstone Ecosystem, including 104 wolves in 11 packs inside Yellowstone National Park. This large and widely distributed population ensures there will be many wolves in the Yellowstone area for people to enjoy into the foreseeable future. However, the shooting of beloved animals or numerous wolves from well-known packs are a public relations nightmare for park managers and state wildlife officials, with advocates insisting on buffers with complete protection around the park and sportsmen retorting with calls for more liberal harvests up to and within the park.

National Park Service policies protect native species and the ecological processes that sustain them, while discouraging human intervention.[91] Thus, the agency has no plans to control numbers of predators such as bears, cougars, or wolves inside Yellowstone National Park. However, the agency has no management authority over wildlife outside the park, where harvests are regulated by state agencies. Instead, park managers have coordinated with wildlife agencies in surrounding states to request smaller harvest units and quotas for wolves near the boundary of Yellowstone. The states have considered these requests and made some changes to limit the harvest of wolves occasionally moving outside the park. Conservative harvests near the park should ensure wolves are available for public viewing which, in turn, protects the regional economic benefits and enjoyment of wolf watching. Harvests should also maintain naturalness and wolf social structure, while not significantly influencing regional hunting opportunities, the control of livestock depredation, or the ability to regulate wolf numbers elsewhere.[92]

The reintroduction of wolves to the Yellowstone area was a transformational event because it completed the restoration of native, large carnivores in the ecosystem, which is a remarkable, though controversial, achievement. The concurrent recovery of bear and

---

[91]National Park Service. 2006. Management policies 2006. U.S. Department of the Interior, Washington, D.C.

[92]Smith, D. W., P. J. White, D. R. Stahler, A. Wydeven, and D. E. Hallac. 2016. Managing wolves in the Yellowstone area: Balancing goals across jurisdictional boundaries. Copy on file at Yellowstone Center for Resources, Yellowstone National Park, Mammoth, Wyoming.

cougar populations,[93, 94] along with human harvests and the uncertainties of weather (e.g., drought, severe winters), facilitated and maintained a substantive decrease in what some people perceived as overabundant elk populations and others perceived as an economic and sporting boon to the region. For better or worse, wolves are generally credited or blamed for substantially depressing elk numbers and other changes to the system that occurred after their reintroduction, including limiting overgrazing by large herbivores (plant eaters), providing food for hundreds of scavengers and decomposers, and redistributing nutrients across the landscape.[95, 96] Certainly, the effects of wolf restoration should not be minimized since they are a dominant, top carnivore whose sociality makes them formidable predators and competitors. However, the reality is that

[93] Haroldson, M. A., F. T. van Manen, and D. D. Bjornlie. 2015. Estimating number of females with cubs. Pages 11-20 in F. T. van Manen, M. A. Haroldson, and S. C. Soileau, editors. Yellowstone grizzly bear investigations: Annual report of the Interagency Grizzly Bear Study Team, 2014. U.S. Geological Survey, Bozeman, Montana.

[94] Ruth, T. K., and P. C. Buotte. 2007. Cougar ecology and cougar-carnivore interactions in Yellowstone National Park. Final Technical Report, Hornocker Wildlife Institute/Wildlife Conservation Society, Bozeman, Montana.

[95] White, P. J., and R. A. Garrott. 2013. Predation. Wolf restoration and the transition of Yellowstone elk. Pages 69-93 in P. J. White, R. A. Garrott, and G. E. Plumb, editors. Yellowstone's wildlife in transition. Harvard University Press, Cambridge, Massachusetts.

[96] Hebblewhite, M., and D. W. Smith. 2010. Wolf community ecology: Ecosystem effects of recovering wolves in Banff and Yellowstone National Parks. Pages 69-120 in M. Musiano, P. Paquet, and L. Boitani, editors. The world of wolves. University of Calgary Press, Calgary, Alberta, Canada.

the overall recovery of the most abundant and diverse predator community in the continental United States is a primary driving force behind all of these changes, not wolves alone.

The recovery of multiple large predators in the Yellowstone area is quite recent (10 to 20 years) in ecological time, and it is unclear how the system will continue to change over subsequent decades as an increasing human presence, warming climate, and these predators and their prey interact across the vast Yellowstone landscape. Obviously, elk numbers have decreased substantially in areas where they were highly vulnerable to predation, with modest effects elsewhere. However, predator numbers necessarily decrease in response to fewer prey animals and, for example, wolf numbers in Yellowstone decreased from about 170 in 2007 to 105 in 2014 as elk numbers declined. Also, elk harvests in nearby areas of adjacent states have been substantially reduced in recent years. Elk should respond to these decreases in predator numbers and harvests with higher survival and recruitment of young. However, the days of 20,000 elk in northern Yellowstone are gone for the foreseeable future.

NPS Photo/Jim Peaco

Essay Nine:

"We're Surrounded, That Simplifies the Problem." Being an Effective Leader

NPS Photo./Neal Herbert

I was probably one of the worst second lieutenants in the history of the United States Marine Corps.[97] Though painful to admit, it was sadly true, as many colonels pointed out at the time. To this day, I marvel at some of the stupid decisions I made. Fortunately, I was a fairly quick study and improved dramatically, eventually grasping the lesson that successful leadership is essentially about two things which are equally important: 1) accomplishing the mission, and 2) taking care of your people. I have maintained this leadership philosophy through many years in wildlife biology, and it has served me well. If you work hard to consistently fulfill these two tenets, you will eventually earn the respect and trust of your co-workers, regardless of what position you occupy in the chain-of-command.

To accomplish the mission, the leader of a particular group or project needs to define the problem or task for their coworkers, answer questions and consider

---

[97] The quote in the essay title has been attributed to Lewis "Chesty" Puller, the most decorated Marine in history.

ideas, refine a plan based on this input, and then supervise the successful completion of the project. Everyone at every level should work at becoming a better leader by participating in the planning of their daily activities, being prepared to accomplish their assigned tasks, focusing and paying attention to detail during these activities and, as necessary, being creative to solve problems. Also, everyone needs to take responsibility for their actions. We all make mistakes—acknowledge them and learn from them. In addition, try to live your life with honor (do what's right), courage (do what's right even when it's hard), and integrity (be true to yourself and others). It's not easy, and I've failed to live up to these ideals many a time, but it's important to strive to attain them.

Successful leadership comes through ideas, planning, supervision, and hard work. The rarest commodity in wildlife biology is an original idea. Wildlife management is often about solving problems, and ideas to reach solutions come from being curious and thoughtful about the world around you. Planning involves setting realistic objectives and timeframes, and preparing for the inevitable contingencies. Detailed planning is essential because, as the Cheshire Cat explained to Alice in Wonderland, any road will get you there if you don't know where you're going.[98] In other words, if you don't define clear objectives and a precise process to attain them, you'll fail to advance toward the desired outcome. Supervision involves choosing good people, training them, providing guidance on what you expect, and then letting them use their abilities and inge-

[98]Carroll, L. 1865. Alice's Adventures in Wonderland. Macmillan, London, England.

nuity to get the job done, all the while monitoring their progress to ensure successful completion.

To take care of their people, leaders need to make sure their coworkers have the proper training and equipment for the task, as well as a strong commitment to safety. When conducting the daily activities, everyone needs to focus on their tasks, be alert to their surroundings and changing conditions, and pay attention to detail. Everyone needs to be prepared to implement contingency actions if things do not go as planned. If anyone observes an unsafe situation, either rectify it or distance yourself and your coworkers from it and then bring it to the attention of your supervisor. If a situation doesn't feel right, back away and assess why. Trust your instincts. Also, if others are exhibiting unsafe behavior, let them know about it and correct it. Don't remain quiet and watch someone get hurt.

Some accidents will happen despite every precaution. The goal is to keep them minor and rare. If someone is injured, make sure they receive treatment and are taken care of. If you are injured, let people know as soon as possible. No one should be punished or ridiculed for reporting an injury, no matter how minor it may appear initially. Also, you should encourage the reporting of close calls or near-misses that highlight a safety concern that needs to be corrected and shared with others to prevent future injuries. Safety needs to be a creed and a culture in your group; people have to believe in it and live it to be successful. For example, the United States military has a creed, never leave anyone behind; that is not always possible or true in reality, but essential nevertheless because everyone believes their buddies will do everything in their power to live up to it. There is no doubt about this commitment; it is absolute and, therefore, reassuring. Obviously, the day-to-day risks taken by wildlife biologists should be far less than our brothers and sisters facing combat in defense of our country and ideals. However, a similar creed is still applicable—everyone should come home safe to their families and friends each day.

Last but not least, I appropriated the following bits of advice and ideas regarding leadership from a wide variety of wise people, ranging from my father, Bert White, to the author of the book Don Quixote, Miguel de Cervantes, to the former Chief of the Yellowstone Center for Resources, John Varley, to famous football coach, Vince Lombardi, to the extremely successful businessman, Bill Gates, to the legendary Marine, Chesty Puller.

• *Try to be Great.* Set high standards, work hard, and make a difference. Also, be passionate—remember why you got into wildlife conservation.

• *Act Like You've Been There Before.* You have to be confident and believe in yourself and your co-workers. Don't let fear or those pesky, naysaying "voices" in your head keep you from acting and succeeding.

• *Don't Fight Windmills.* Focus on the important things you can change, not the things you can't. Also, if you try one approach to solving a problem and it doesn't work, repetition likely won't help. Think, listen to the ideas of others, and come up with an alternate solution.

• ***Don't Be a Sensitivity Kitty.*** Expect criticism; when people stop criticizing you, they've given up on you. If you need to impart criticism, do it soon after the mistake is made and the events are fresh in everyone's mind.

• ***Don't Be a Talk Show Host.*** If it's not helpful, don't say it or spend time on it. Also, listen to people instead of thinking about what you're going to say next. Keep an open mind, and compare and contrast alternative actions before making a decision.

• ***Just Do It.*** Once the boss makes a decision, you don't have to like it but you do have to implement it unless it's unethical, unlawful, or unsafe. Work is not a democracy. Likewise, you don't have to like your co-workers, but you do have to learn to work effectively with them and take care of them.

• ***Don't be a Hero.*** Don't try to do it all yourself. Surround yourself with smart, talented, hard-working people and learn to delegate. Believe it or not, some other people are better at certain things than you are.

• ***Success Breeds Success.*** Have pride in your accomplishments, and encourage and reward others for excellence. Put people in a position to succeed, and use their strengths and talents to boost productivity and confidence.

• ***Take Things in Stride.*** Sometimes, despite your best planning and efforts, everything goes to hell for whatever reason; and you can either sit down and cry or work your way out of it. Life's not fair—so get over it and deal with the problem(s). As we said in the Marine Corps when things went from bad to worse, "ain't nothin' but a thing."

• *Play to Win.* People are paying you to accomplish great things, not just to show up for work. Help yourself by learning to take calculated risks, when appropriate, and negotiate effectively, whether the issue is time, power, money, or some other matter.

• *You Can't Have Too Much Fun.* Stop taking yourself so seriously and enjoy your job. Relax and have fun with your colleagues—which is much easier if you're working hard, solving problems, and winning.

These teachings have been very helpful to me throughout my wildlife career, and I hope they will be helpful to you during your journey. And remember, no one will help you if you don't first help yourself by putting forth maximum effort.

NPS Photo/Jim Peaco

Essay Ten:

Tea with the Mad Hatter: Negotiating the Impossible with the Irrational

NPS Photo/ Jim Peaco

Long ago, in what now seems like a previous life, I took a job with the Fish and Wildlife Service in southern California.[99] My task was to conserve dozens of endangered and threatened species in the face of rampant and inevitable human development. Imagine trying to conserve remnant populations of beetles, birds, butterflies, kangaroo rats, and other animals when they live on land worth a tremendous amount of money for development. My first project, before I'd even received training on the Endangered Species Act, involved a massive water development and transfer project. I'll never forget walking into a conference room for the first meeting and sitting down across from a cadre of biologists, lawyers, and project managers from the company proposing the development. They proceeded to bury me with reams of paperwork and analyses,

---

[99] The reference to the Mad Hatter in the essay title is from Lewis Carroll's 1865 book Alice's Adventures in Wonderland published by Macmillan in London, England.

and tell me in no uncertain terms how and when the project was going to proceed. When they were done, I not-so-politely informed them they were sadly mistaken. From there, things quickly deteriorated into a shouting match with threats hurled every which way.

This fledgling introduction to the world of negotiation was enlightening to say the least. I never had a class or instruction on the topic and, as a result, was pathetically unprepared. Given this predicament, I rushed to find information to save my ass and, by dumb luck, stumbled upon the teachings of Herb Cohen.[100] Everything in this essay that is worthwhile stems from his wisdom. To paraphrase Mr. Cohen, negotiation is about obtaining, evaluating, and using information, power, and time to change people's behavior and satisfy mutual needs, thereby attaining your desired outcomes. Negotiation is a predominant part of wildlife conservation, whether you are interacting with your boss, private landowners, developers, business owners, or people from other federal and state agencies, tribal governments, or non-governmental organizations.

For example, the preservation of wildlife that move across the boundary of Yellowstone National Park and are exposed to different jurisdictions and management paradigms throughout the year requires close coordination and collaboration with other federal and state agencies, private landowners, and local governments. Overall, federal and state agencies have similar goals to manage sustainable populations of wildlife as public resources for the benefit of people. However, the

---

[100]Cohen, H. 1982. You can negotiate anything. Random House Publishing Group, New York, New York.

somewhat different focuses of these agencies can, at times, make it difficult to reach consensus and lead to quarrels about issues such as the appropriate numbers of certain species, potential for disease transfer from wildlife to livestock, and the ecological and tourism value of predators versus their effects on livestock and game species such as deer and elk. These issues have long-standing, adversarial histories, with skirmish lines drawn and fortified over time. Therefore, biologists and managers need at least modest negotiating skills to quell the initial frustration and distrust, find common ground, and work collaboratively to solve problems and reach a workable solution for everyone.

Even after years of effort, I am not a great negotiator. Ignoring many of Mr. Cohen's warnings, I tend to care too much about the outcome and, as a result, can become quite aggressive and confrontational. I curse like a Marine (sailors are rookies) and have been called a bully on more than one occasion. Despite these faults, I have been successful at negotiating many complex, contentious, and important deals during my wildlife career. These successes have come from trying and adapting the advice conveyed by Mr. Cohen to my own abilities and personality. In the following paragraphs, I've attempted to paraphrase many of Mr. Cohen's main precepts, incorporated with bits of advice based on my own experience.

Be Prepared. You should always strive to have better information than your counterpart. So spend the time to get it, and think about how and when to use it to your advantage. Before the first meeting, analyze areas of potential disagreement and their likely causes.

Then come to the negotiating table with feasible alternatives that could satisfy the needs of all parties. It's advantageous to you when your counterpart agrees to discuss and pursue one or more of these options. At each meeting, convey confidence and demonstrate you have the ability and authority to get things done in a timely and efficient manner. Show your commitment to a successful outcome by sharing information and ideas, and striving to reach the best result for everyone.

Know Yourself and Your Counterpart. Know your strengths, weaknesses, and tendencies—not what you want them to be, but what they actually are. If you don't know them, I'm sure there are plenty of people that will happily share them with you. Know what you want (the desired outcome) and what you're willing to accept (the bottom line) if you can't get there. Try to figure out the same limits for your counterpart, including their deadlines, needs, and pressures. Use this information to anticipate their tactics and adapt your style to more effectively work together towards solving real and perceived problems. Make sure your counterpart has the authority to make decisions. If not, get the right person there.

Identify the Real Needs. Winning a negotiation is not about conquering, but gaining enough trust and commitment from your counterpart to facilitate working together to solve problems and achieve solutions. Thus, you have to uncover the real needs of your counterpart, which can be difficult if they do not articulate them, which often occurs if you're meeting in a public venue. To identify their real needs, listen to their demands and analyze what they're saying. If necessary, get them in an environment where they'll talk more openly. Once you know their real needs, try to look at things from their perspective and develop solutions that satisfy them in addition to your own needs.

Be Your Self. Use your unique abilities and talents to persuasively promote a solution where both parties can get what they want. Don't try to emulate someone else or manipulate anyone. Instead, identify mutual interests with your counterpart and work from there to reach solutions. Don't get into the weeds—at least not at first. Reach agreement on the big picture before delving into minutia. If you're not making progress on a tough issue, shelve it for a while and discuss other matters. If possible, use examples, precedents, and perhaps even the opinions of important stakeholders to justify your ideas. Don't be self-righteous about your mission and needs, after all your counterpart has similar expectations and pressures from bosses, politicians, and stakeholders. So show some understanding and respect.

Make it Personal. Reaching a good deal is all about the relationship between you and your counterpart. Try to establish commitment, respect, and trust. Be true to your word and don't divulge information to third parties. If you make a concession during the negotiation, get something in return, even if it's just an acknowledgement of your overwhelming sacrifice. If your counterpart becomes adversarial, politely let them know it would be unwise to break your trust—but don't give them any inkling of the specific consequences. Let them fret. Conversely, your counterpart needs to believe you'll agree to a good deal for them if they negotiate in good faith. To lessen conflict, try changing the way you address, react, or respond to your counterpart in an attempt to change their behavior. It may or may not work.

Expect and Counteract Opposition. There is always some resistance to change, so expect it and deal with it. Dissent can help you develop and fine-tune a collaborative solution where both party's needs are satisfied. Envision what challenges may arise, and prepare for them. This preparation will help you avoid making emotional or reflexive reactions. Make your opening gambit a good one—collaborative, convincing, and rational. Be frank and honest in addressing naysayers, without being disparaging or intimidating. If you need to debunk misinformation or misperceptions, do so with kindness, questions, and evidence. Also, be prepared to react to counteroffers with conscience and common sense. Avoid arguments and jibes that harden positions, and destroy progress and trust. And never attack someone personally in public unless you want an enemy for life. Instead, use your preparation and talents to lead them towards a successful solution.

If Necessary, Take a Hike. If your counterpart is adversarial, they may initially make extreme demands to elicit concessions from you. Don't reward them by trending towards the middle between both party's positions to reach a compromise that satisfies no one. Instead, get people talking about the issues so you can figure out your counterpart's real needs and present effective ideas to solve problems, overcome disagreements, and reach a collaborative solution that works for both parties. You may need to go back and forth through lots of discussions to reach the moment where you can attain the outcome you want. Also, you have to be willing to leave the negotiating table and wait for a better time if demands, politics, or tensions are too high. Don't rush in response to derived timelines or ultimatums. Instead, consider your options and make wise decisions based on calculated risks.

It is not easy to reach a collaborative solution that satisfies the needs of both parties. In fact, it is often very difficult and time consuming. Sometimes, you have no choice but to compromise in the interim to keep things going in the field, while you continue to work toward a better long-term deal. However, try to keep your eyes on the prize and not be satisfied with compromise. By adapting Mr. Cohen's advice to your own unique personality, style, and talents, most of the time you can eventually reach a successful, collaborative deal. As Mr. Cohen has pointed out, your bosses and peers may even acknowledge your brilliance for an hour or two.[101]

[101] Cohen, H. 1982. You can negotiate anything. Random House Publishing Group, New York, New York.

FINALE

In closing, I'd like to reemphasize several key points. Effective wildlife conservation and management is largely about generating innovative ideas and practical options, developing effective coalitions, and collaboratively solving problems.[102] Many of your initial assumptions and perceptions about the conservation and management of wildlife, as well as the cultural and social values of people in various agencies and stakeholder groups, will be challenged over time. As a result, frequent assessments of the assumptions, expectations, and values held by yourself, your colleagues, and other interested parties will be essential for progress. Learn from your experiences, share the lessons with your colleagues, listen to the concerns and ideas of various stakeholders, and adapt management activities to better progress towards desired conditions. The task of conserving and managing wild, wide-ranging

[102]Clark, S. G. 2002. The policy process: A practical guide for natural resource professionals. Yale University Press, New Haven, Connecticut.

animals across a vast and unfenced landscape, while minimizing conflicts with humans, is a difficult, but critical, endeavor in modern society.

In addition, I'd like to offer a few more bits of unsolicited advice. First, we need to remember the "faces of our fathers,"[103] by which I mean the contributions and insights of our predecessors. Not long after arriving at Yellowstone National Park, I realized previous biologists and managers had dealt with many of the same issues for decades and in doing so, developed innovative approaches for conserving wildlife in preserves that are still emulated around the world. A few examples include decisions to minimize human disturbance and allow natural processes to prevail, "rewilding" food-conditioned bears, restoring other large predators such as wolves, and returning the role of fire as a natural disturbance process in the ecosystem. The men and women who took these bold actions, which were all widely criticized and denounced at the time, deserve our admiration and respect for their guidance, inspiration, and leadership.

Second, I often hear grumblings from my veteran and wildlife biologist buddies about how young folks today just don't measure up to those of yore. Don't listen to these biased recollections that are probably just a manifestation of mad cow disease anyways. Having grown up in yore, I can assure you that today's biologists are smarter, better educated and trained, work just as hard

---

[103] Paraphrased from Steven King's book The Dark Tower: The Gunslinger, which was first published in 1982 by Donald M. Grant, Publisher, Inc., Hampton Falls, New Hampshire.

and, when you hide their cell phones, focus just as well. Thus, the future of wildlife biology is bright and in good hands.

Finally, make the time to laugh with your friends. One of the most beautiful and contagious things I've experienced is the uncontrolled and uninhibited laughter of a child. It is so contagious that you cannot help but join in, even if you have no idea what the joy is about. We all need similar moments in our lives and, as a result, you can never have too many good times with good friends.

Thanks for listening. Semper fidelis![104]

[104] This Latin phrase, which means "always faithful," is the motto of the United States Marine Corps.

ACKNOWLEDGEMENTS

I dedicate this book to my dad who was my hero and gave me the drive and wit to succeed when everyone else told me I wouldn't; my mom who gave me the feistiness and passion to venture into the unknown and experience life to the fullest; my sister who is the kindest person I know and has selflessly taught generations of teenagers for more than 30 years; my brother who became the father and good man that I strived, but failed, to be; and my son who, despite my parenting blunders, has grown up to be a fine young man and makes me extremely proud.

Now that my son has made it through school, I'll confess to him a few fibs. I didn't actually get all A's in school; we actually did lose a few football games; the walk to and from school wasn't uphill both ways; and occasionally the snow we had to shovel was less than 10 feet deep.

I thank Jennifer Carpenter, Susan Clark, Bob Garrott, Dave Hallac, and Dan Wenk for reviewing the essays and offering much needed advice and edits. Bob Garrott went the extra mile and edited the entire manuscript, for which I am extremely grateful. Bob and his wife Diane have been my best friends through my wildlife career, always there to provide mentoring and support, and even bailing me out of jail on occasion.

I thank Pat Bigelow, Chris Geremia, Kerry Gunther, Todd Koel, Doug Smith, Dan Stahler, John Treanor, and Rick Wallen for sharing their thoughts and ideas on several essays in this book and, more importantly, for their endless dedication and hard work toward preserving natural resources in Yellowstone.

I thank Julie York for her unbelievable kindness and generosity in contributing to my son's education.

Last, but certainly not least, I thank my friends from the village of Candor in upstate New York where I grew up. There's no way I would have made it without your encouragement and support. Many thanks for keeping most of the embarrassing stories about me to yourselves.

NPS Photo/Neal Herbert

P. J. White is the Branch Chief of Wildlife and Aquatic Resources at Yellowstone National Park. He received the Director's Award for Natural Resource Management in the National Park Service during 2010. He has collaborated to produce three other books on Yellowstone, including The Ecology of Large Mammals in Central Yellowstone: Sixteen Years of Integrated Field Studies (2009; ISBN—13:978-0-12-374174-5); Yellowstone's Wildlife in Transition (2013; ISBN 978-0-674-07318-0); and Yellowstone Bison—Conserving an American Icon in Modern Society (2015; ISBN 978-0-934948-30-2). P. J. received his doctoral degree in Wildlife Ecology from the University of Wisconsin (1996); master's degree in Wildlife Conservation from the University of Minnesota (1990); and bachelor's degree in Wildlife Science from Cornell University (1980).